YEAR OF THE DUNK

ASHER PRICE

YEAR

of the

DUNK

A MODEST DEFIANCE
OF GRAVITY

CROWN PUBLISHERS
NEW YORK

Library of Congress Cataloging-in-Publication Data
Price, Asher.
   Year of the dunk : a modest defiance of gravity / by Asher Price.—First Edition.
     pages   cm
   Includes bibliographical references and index.
   1. Dunking (Basketball)   2. Basketball—Offense.   I. Title.
   GV888.15.P74 2015
   796.323'2—dc23   2014041716

ISBN 978-0-8041-3803-1
eBook ISBN 978-0-8041-3804-8

Printed in the United States of America

Jacket design: Rodrigo Corral
Jacket photographs: Susanna Price / Getty Images (basketball hoop);
   Jaroslaw Wojcik / Getty Images (player)

10 9 8 7 6 5 4 3 2 1

First Edition

For Bubu

"Mama exhorted her children at every opportunity to 'jump at de sun.' We might not land on the sun, but at least we would get off the ground."

—ZORA NEALE HURSTON, *Dust Tracks on a Road*

Jack be nimble, Jack be quick.

Jack jump over the candlestick.

Contents

CONTENTS

YEAR OF THE DUNK

Introduction

Nineteen eighty-six: a fun-sized Superman with close-shaven hair, dimples, and an adorable potato-chip of a name floats in his short shorts toward a basket.

It is the finals of the National Basketball Association slam-dunk contest at Reunion Arena in Dallas. Before a crackling Saturday night sellout crowd of 16,573, in midair, is Spud, the unlikely corruption of the nickname Sputnik, earned by Anthony Webb as an infant, not for some obvious early ability to launch himself skyward but for his unusually large head. A native son—his parents own a convenience store in black South Dallas—he knows the scalpers who sold him the tickets he needed tonight to pack in his three sisters, two brothers, and mother and father. A good thing, too—he got a hometown discount.

The yolk-orange rim, like those on all official baskets, is 10 feet off the ground. Webb, at his perkiest, stands five feet seven inches tall: no taller than a parking meter, as one newspaper commentator has described him. In fact, arena security has turned him away more than once when he's reported to road-game locker rooms. He weighs 133 pounds and can't even palm a basketball. But his legs— they are the thickness of bowling balls.

A 22-year-old rookie, Webb finds himself squaring off against his far more famous Atlanta Hawks teammate Dominique Wilkins. Also known as the Human Highlight Film, he is, at 6'8" and 224 pounds, the defending slam-dunk champion. Webb makes the league minimum, $70,000; Wilkins, $585,000. In regulation play, Webb normally feeds Wilkins the ball, yet on this early February night in Dallas the pair is trading acrobatics around the hoop, throwing down one jam after the other in a show of skywalking one-upmanship. The winner gets nearly a fifth of Webb's annual salary: $12,500.

The judging panel is made up of several retired NBA players, former Dallas Cowboys quarterback Roger Staubach, and, curiously, Martina Navratilova. Webb has dunked many times before—as far back as high school, at a height of only 5'3". In a sense, Webb's lightness works in his favor: Putting aside paper planes and Wiffle balls (disqualified for their aerodynamic flaws), you can throw a lighter object farther than a heavier one. Force equals mass times acceleration, and Webb has less mass to carry skyward. But to lift his 5'7" frame to the rim, he must jump an extraordinary 42 inches off the ground, higher than your kitchen sink. And he faces the obvious physics problem encountered by any jumper: The moment your feet lose contact with the ground, you have no additional force to exert—even as the force of gravity is pulling you back to earth. His solution to all these problems is elegant in its ferocity: He gathers and applies all his muscle strength in the shortest time possible, about a tenth of a second—the time his foot plants before he shoves off toward the rim. In essence, little Spud Webb is exerting nearly four Gs to push off the ground, about the same acceleration a fighter jet creates as it blasts off an aircraft carrier.

An avalanche of dunks. Spud, small, self-contained, starts loud, with a reverse slam he throws down so hard that the ball ricochets off his still-airborne head and flies back up through the basket, as if to announce that, yes, indeed, I can put it down. From there, it gets

2

increasingly fancy: a short run-up to a two-handed double-pump dunk; a 360-degree helicopter one-handed dunk (a.k.a. "the Statue of Liberty"); a lob pass to no one that bounces high off the hardwood before Spud, in one fell swoop, catches it, spins 180 degrees, and jams—a nasty bit of self-dealing. Each dunk begins with the run-up, then a super-brief foot-plant in which the legs coil and the Achilles tendons stretch to store up accumulated potential energy—and then, *wham!* Unleashed, upward kinetic force. To launch himself into orbit, Webb spends that coiled-up phase pushing against the hardwood floor with all those Gs, moving from zero miles per hour at the plant of a two-footed dunk to roughly 10 miles per hour through the air in less than a quarter second.

Countering, Wilkins executes some of the same muscular dunks that won him the contest the previous year: First a two-handed windmill dunk, in which the ball is spun around as if it's in a washing machine before being thrown through the hoop; he manages, next, a dunk that starts as a windmill and ends as a one-handed tomahawk, the fierce piercing of the basket that not even vaguely recalls the American Indians; and then, beautifully, he performs a reverse dunk in which he reaches the ball down toward his ankles even as he is ascending through the Reunion Arena ether and then, swiftly, pulls it back and slams it behind his head just before making his way back to planet Earth.

A TV man asks Martina what she makes of the contest thus far. "In my next life," she says, "I'd like to come back as a black basketball player."

Webb has one final opportunity to best Wilkins. The crowd, now clearly in the corner of the little guy even as it respects the Highlight Film, starts chanting "Spud." He moves to half-court, his lightly upraised fist moving in small circles, Arsenio Hall–style, as he prepares for the coup de grâce: a one-handed overhand bounce pass that leaps from the ground, bounces off the backboard, and goes back into his outstretched, soaring hand. He snatches the rock as a quick, collective inhale whooshes through the arena, and slingshots it home.

Martina goes nuts. Staubach goes nuts. America goes bananas.

And, as a six-year-old in New York City, I looked up from my GI Joes toward our bulky Sony television to witness the mini miracle. "Let's go to the videotape!" shouted Warner Wolf, the WCBS sports-caster, and suddenly seeing the small, boyish man—only a few inches taller than I was!—dunk, I was captivated. In the odd pick-and-choose cultural moments imprinted upon a kid's brain, Spud Webb joined a constellation that burned with the faces of Christa McAuliffe, the teacher chosen to board the *Challenger* space shuttle, which less than a month earlier had disintegrated on its way toward outer space; Ronald Reagan; the entire roster of the New York Mets, who would go on to win the World Series in October; and Billy Idol, whose "White Wedding" music video my two older brothers obses-sively watched on MTV. That, for me, was the sum of 1986.

Webb himself didn't know how to describe his unlikely ability.

"I haven't the slightest idea," he said when asked about how he jumped so high. "When I find out I'm going to write a book about it. I guess it's just God-given talent."

•

This is the book that Webb never wrote. Maybe Spud Webb's gifts were God-given; on the other hand, maybe he somehow bested the destiny appointed to his five feet, seven inches. I'm interested in the limits of human talent—why some people have the mojo and others don't, and how far, as individuals and as a species, we can push what talent we possess.

Everybody wants to dunk, at least metaphorically. We think that if we spent just a year away from our everyday distractions, we could rise above our terrestrial lot: learn Spanish, pick up the piano, remaster calculus, paint. In our fantasies, we think we might all be naturals—the capability of mastering some talent hidden inside us. A few years ago, *The Onion* cheekily mocked our unspent dreams in an obituary with the headline "97-Year-Old Dies Unaware of Being Violin Prodigy."

The notion of a "hidden talent" can haunt, too. My mother stopped making art after a junior high school teacher told her she had little talent; she became an art historian instead, her days spent tromping through museums to examine other people's work. It's a familiar story: We leave our singing in the shower. Most adults never bother to pick up a violin, write fiction, or learn other languages. Why acquire a talent just to explore its limits?

I meant to take this dunking metaphor literally: I wanted to slam a basketball through an orange rim. My quest was to make the most of the piece of flesh I'd been given. At the extreme margin of human talent and effort, elite athletes stretch the boundaries that define our capabilities as a species. Will there come a day, the former Trinidadian sprinter Ato Boldon was once asked, when someone runs the 100-meter dash in less than nine seconds? (The record is now 9.58 seconds, set by Usain Bolt.) "Sprinters believe that—someday—somebody will run the 100 meters and the clock will read 0.00," Boldon said. "And when a sprinter thinks like that, he's not trying to trick himself. It's how you have to think. This idea of human limitation is exactly what we're competing against." I would never run as fast as Bolt or Boldon. It's just not in my DNA. But the test I had set myself was just possibly manageable: Given my height and vague athleticism, I felt that with a lot of effort I should be able to push a nine-and-a-half-inch ball through a 10-foot-high hoop.

I faced some challenges. I'm of Austro-Hungarian stock, more closely associated with making good pastries than with jumping ability: At the start of this project I could only swipe the rim with the tips of my middle and index fingers. As Sidney Deane (Wesley Snipes) famously tells Billy Hoyle (Woody Harrelson): "Billy, listen to me: White men can't jump." I owned healthy love-handles—I weighed 203 pounds—so I was going to have to lose weight and put on muscle. But I had some things going for me: height—I'm 6'2½" with orangutan arms; what a former coworker once called a "big ol' sprinter's butt," just the kind of powerful posterior I'd need to propel myself hoopward; and, as I neared my 34th birthday, some leftover sportiness (I had never played a varsity sport, but once upon a time I had captained my college Frisbee team). I had never weight-lifted, either—I despise weightlifting—and so, to my mind, at least, I remained a tabula rasa. "Pure potential," my wife, Rebecca, said, with a not-so-small degree of skepticism.

Naturally this would be a navel-gazing exercise: "I should not talk so much about myself if there were anybody else whom I knew as well," Thoreau wrote, by way of apology, at the outset of *Walden*. Because in its bones this is a peculiarly American story, a story about optimism, about self-reliance, about the ability to remake oneself, Thoreau, who counted on the labor of his two hands just as I was to count on the labor of my feet, was a lodestar. The dunk is, yes, as American as jazz or apple pie. But it stretches beyond that—it is literally about upward mobility, about the very American idea, evoked so often in *Walden* and by Thoreau's contemporaries, that everyone is capable of self-improvement, of rising above her lot. For me, the test was physical; for others, the barriers, involving everything from class to gender, are obviously harder to overcome. Americans have long thought that they could move on up, as they say in *The Jeffersons*. They believe in self-made men, and, I suppose, that's what I was trying to do: remake myself.

On some level, then, this mission is also about wish-fulfillment.

In 2008, candidate Barack Obama was asked whether he'd rather be the president or Julius Erving, the great dunker of the 1970s and early 1980s, in his prime. "The Doctor," he said like a shot. "I think any kid growing up, if you got a chance to throw down the ball from the free-throw line, that's better than just about anything." (Obama first dunked when he was 16: "I still remember the day that it first happened," he told *Sports Illustrated*. "One of those magical days when you're just in a zone.")

Yet there is an unfairness to our birthright: You grow up and the fantasy fades. Every kid wants to be an astronaut or an NBA star, but pretty quickly as we grow older we're narrowed down into doing something more realistic. Some people simply aren't as good, can't be as good, as others. Basketball, like other sports, illuminates how far effort will take you. LeBron James surely works his butt off, but he was also just born with more capital. And the dunk, shorn of the thousand techniques and instincts that take years to develop on a court or a field, further crystallizes that difference: It is a simple, extremely basic maneuver; for my purposes, the beauty of the dunk (as a goal) is that either you can do it or you can't. It has empirical finality.

Hovering over my story is the psychological question of how we disappoint ourselves when we grow up and how we try to overcome that disappointment. A "slam-dunk," of course, means a "sure thing," and in this sense the dunk stands counter to the adult challenges that we face, finding certainty in our lives amid the shifting questions of career, of marriage, and of children. Trying to dunk appeals because there is something childlike and uncomplicated about it. And going after it was a way, in a modestly rebellious act, of not acting my age. The idea of an adult trying to learn to dunk seemed half-ridiculous, as if trying to re-create the kind of footloose physical freedom of childhood. Yet that's part of the joy of sports: Someone once asked Michael Johnson, the great sprinter, what it was like to run as fast as he did. He said the only thing that compared was, as a kid, going downhill go-karting on his birthday. "So you go get yourself a go-kart and find yourself a hill, and you'll know how it feels."

I was no kid, making this my last chance to dunk. I gave myself from the end of one August to the end of the next to improve. It was a year to discover whether, embedded in my bones, muscles, and DNA, was some grand jumping potential. A year was long enough to train my feet, hips, legs, and butt in the strange arts of explosive movement. Any longer and I figured I'd see diminishing returns; competing against my efforts were the natural force of aging and the reality of having to return to my daily work. Besides, working out the way I did was a wretched business, and I'm not one to engage in too much voluntary misery. The only rule I set myself was what I'll call the no-needle pledge—I wouldn't ingest or inject any medical-grade material to make myself jump higher. Lord knows I'd absorbed enough of that stuff in my time.

●

In many ways, I'm Spud's opposite. I grew up on the soft-scrabble streets of Manhattan's Upper West Side, on a steady diet of whitefish schmear and bagels. I had shaped up as a thwacker of awnings. After Hebrew school every Tuesday and Thursday night, delighted to be liberated of religious instruction, I danced home on the sidewalks of 86th Street, and, in stride—hop, hop, leap—swung my left hand up and against the overhead canopies. Grouchy doormen gave me the stink-eye, but by the age of 12 I could touch at least half of them. On the highest ones I could just rattle the hidden metal bar that stretched the awnings taut and, for the lower ones, drum the canvas awning itself. Fine black soot powdered my fingertips. The awnings were spaced regularly, like ships huddled against piers. I had a kind of running rhythm, like overgrown hopscotch, making sure to get on my right foot for takeoff at the awnings I wanted to smack. Swinging shop signs and construction scaffolding brought their own satisfaction. Just make contact. I was like a Little Leaguer of the streets.

Hitting the lowest of these canopies, each maybe eight feet off the ground, wasn't too hard: At that age I was always second-tallest in school. The tallest was Danny Rosen, a friendly, dopey-faced giant with the splayed feet of a duck and the sort of perpetual tan that announced that he had wintered in Vail and weekended in the Hamptons. Danny couldn't jump. I wasn't much of a hopper, either, but what I lacked in ups I made up for in enthusiasm: Like so many Jews, I never set foot in Hebrew school after 13, the age of Bar Mitz-vah, but I kept lifting off, like a boy, at the sight of an awning. In my teenage years, hanging out with friends, especially girls, it was a silly, boyish way to show off. I leapt and leapt and leapt.

But there was a natural ceiling to my awning-slapping. I never could reach the loftiest ones. I didn't really try, either. I was one of Frost's swingers of birches, as interested in coming back to earth as leaving it:

> He always kept his poise
> To the top branches, climbing carefully
> With the same pains you use to fill a cup
> Up to the brim, and even above the brim.
> Then he flung outward, feet first, with a swish,
> Kicking his way down through the air to the ground.

I happily smacked the awnings I knew I could smack, and didn't bother with the ones I knew were just out of reach. Happy-go-lucky, you could say, but luck gets you only so high.

Now I am all grown up, and, except for the times when no one, save my wife or an easily impressed young nephew, was looking, I haven't tried to take a whack at an awning in a long time. It's a little embarrassing for a man to hop about Manhattan trying to touch signs. But I have wondered just how high I could get if I really worked at it. If I made it my single purpose for a year—if I spent hours doing leg work and jumping exercises, consulting with basketball trainers and scientists, and talking to top-notch athletes— could I get high enough to dunk?

PART I
the RUN-UP

1

Assembling the Gurus

On a late winter afternoon in New York at some basketball courts by the Great Lawn in Central Park, my hands jammed into my hoodie pockets, I waited for my pal Nathaniel. It was crisp, still cold enough to see the breath puff in front of your face, especially if you were winded. An old Spalding street ball, circa 1988, dug out from my childhood closet, sat on the ground between my feet. In its way, it was a worn-down globe of high-impact asphalt geography. In some parts, the leather stuff, or whatever it once was, had completely worn away, making what remained look like the splitting of continents. Between them, you could see the crisscrossing rubber filaments left naked, like open ocean.

I had asked Nathaniel to meet me because I wanted to entertain him with the latest in a recent line of mischievous ideas. I had struck a restless mood, the sort that comes naturally to a reporter used to facing daily deadlines and now facing open time. My home and my newspaper were in Texas, but I was up in New York on a months-long journalism fellowship, with little more to do than take a few classes, and found myself dreaming up, toying with, and then, inevitably, discarding a succession of playful projects. A history of Jews eating pork. A documentary about a day in the life of an American supermarket. An international comparative study of the milk shake.

I gathered myself up when Nathaniel arrived, all curly hair and glasses. I wanted to see just how far away I was from dunking, I told him. He's a psychiatrist, and, in his practiced way, he nodded as if my crazy idea sounded entirely sensible. My friend since grade school, he has long been a coconspirator in my fantasies, dating to the massive battles we waged with toy soldiers. As Nathaniel stood aside, a plaid scarf swept around his neck, I tried jumping rimward a couple of times with the ball and had my usual pathetic results. An observer would have justly wondered why I was grunting so hard to undertake what appeared to be mere layups. I felt bloated, and the hair on my head seemed to me especially wispy and thinning. What the hell kind of idea is this? I wondered.

Just then, a guy wearing a Megadeth T-shirt and a purple bandanna and Lehigh gym shorts joined us on the court. He was stocky, and there was something formidable about his forehead. He was maybe my height, but he didn't have my long arms. Nathaniel and I were chatting, and this guy asked if he could borrow my ball. As he shot around I asked him if he could dunk. I didn't expect much, but Megadeth just went ahead and did it: threw it down with one hand after a quick dribble and jump. He looked like a flying caveman, with a basketball instead of a club. Then he watched me try, a few times, and gave me some pointers on my biomechanics. I was leaning too far forward. I ought to bring my butt farther down before springing up. I should try jumping off two feet, he told me. As if my inherent failings could all be corrected.

His name was Tyler Drake, and it turned out he had started dunking only at age 30, after an old roommate who had become obsessed with driving golf balls as far as possible began swearing by squat lifts to improve his power. "My college basketball team didn't emphasize weightlifting," he told me. What sort of college basketball team doesn't emphasize weightlifting? And had a guy on the roster who couldn't really dunk? Caltech, it turns out, where Tyler studied physics. When Tyler played on the team, the Beavers (nature's engineers) had not won an NCAA game since 1996. Only in 2007, several years after he graduated, did the team snap the streak—after a record 207 consecutive losses against other NCAA Division III teams. I told him I wanted to make myself a dunker. "Definitely doable," he said. Will you help me? I asked. (Why not? I figured. Don't blink when serendipity stares you in the eye.) He was not a man to dance around words: "Totally." Perfect. A dude who played on the losingest college basketball team ever would give me advice on how to dunk.

A few weeks later, over wine and homemade baba ghannoush at his girlfriend's place in Harlem, Tyler began going over the physics of the dunk. "Right now you're just standing, exerting two hundred pounds on the ground. Jumping, you want to exert five hundred," he said. I scribbled notes with one hand while eating with the other. I tried not to stare at his beautiful girlfriend, Paola, a Haitian émigré and model. A photo of her, naked, with a dusting of sand—but artistic, you know?—hung on a wall. The key to upping ups, he said, was in the routine, obsessive exercise of lifting weights through squats. "The bar is a stronger version of gravity. From a state of motionlessness, you want to go directly upward, and directly downward. The bar pushes you toward perfect form. If you're leaning backward and pushing up, at an angle at all different from exactly up, you're going to eat shit. It demands pushing exactly up. It's totally what you want if you want to jump higher."

Tyler told me he was able to dunk when he could squat 315 pounds and deadlift 405. "It's like loading up a spring and then going up," he said.

"Sure," I said, with a hint of false swagger. I had no idea what that truly meant—I had never really lifted weights—but at least I had something to aim for.

Then he told me that I'd probably develop a line of calluses along my back, from one shoulder blade to another, which, to be honest, kind of grossed me out. But to Tyler I just smiled. I didn't want him to think I would be anything less than fully committed.

●

To transform my average-Joe body into a svelte jumping machine, I had decided to get some professional help. If I was truly going to test my capabilities and scientifically go about monitoring my performance, I needed some top shelf trainers on board.

Tucked on the ground floor inside an expensive-looking condo tower called the Bellaire, the Performance Lab of the Hospital for Special Surgery looks a little like an overgrown, high-end preschool playroom: Large, brightly colored inflatable balls sit on the softly matted ground. Stacks of stepping blocks stand in a corner. Dowels lean against a wall. Fluorescent lights buzz overhead. This is where the New York Knicks have been put through their paces. And now, somewhat amazingly, this was where my mission began.

I hadn't expected anyone to take me seriously. I wanted to dunk, yes, but the notion of an adult going deliberately about it seemed ridiculous. I wrote emails: "I'm interested in the limits of human potential," I explained, as I laid out my project. And then: "I want to dunk." Many of these, unsurprisingly, went unanswered.

And then a real-life scientist, a very nice older man named Stephen Doty, who had himself played high school basketball in the 1950s, wrote me back. "What a great idea for a story. I, too, only reached the rim, never over it. And when I see what these giants of today are doing, it makes me wish I had tried harder."

A couple of weeks later, I found myself hurriedly buying some chocolates at a deli as I hustled to a swank Upper East Side neighborhood to meet Doty, a senior scientist who specialized in the loss of bone density at the Hospital for Special Surgery. He introduced me to a crew of physical trainers who had gamely agreed to help me.

"So we're going to get you to dunk, right?" Polly de Mille, a handsome, sandy-haired reed of a woman said after greeting me with a warm, Massachusetts-sounding hello. She's not the chocolate-eating variety, I thought, immediately regretful that I hadn't bought something like a bag of clementines. From my emails introducing my quixotic project, I had become something of a curiosity, and this was the first time the staff at the Performance Lab had set eyes on me. "That's the idea," I said, trying to sound as confident as I could, despite my own doubts—about my abilities, about my willingness, about my free time.

Over nearly three hours, Polly put me through a battery of exercises to test my capabilities. Or "deficiencies," as one of her colleagues described them. As it would turn out, I'm quite deficient.

She began by taking baseline measurements of my body, an accounting of my flesh-and-blood vital statistics. First she performed a skinfold test: I felt a little like a piece of meat on a hook as she measured the body fat around my chest, arms, and legs with a pair of calipers. The pinching of the skin was meant to determine the subcutaneous fat layer—the measurements are then converted to body-fat percentage. (To determine the mathematical formula, doctors turned long ago to cadavers: After submitting the dead to skinfold tests, they dissected them to determine fat content.) Her findings: About a fifth of my overall weight was body fat, just worse than average for men my age. Between my love handles were the modest beginnings of a gut. Polly, who wore a T-shirt that said "Train Like a Knick," told me that basketball players have body-fat percentages of only 10 percent. To reach that goal, and achieve the physique of a prototypical dunker, at my current level of fat-free mass, I would have to cut my overall weight to 171.6.

And it became fast apparent I would have to add some serious muscle. My brother-in-law Daniel said he thought I would want to get the largest legs possible, and then trim as much as I could from everything above the hips. "Don't you want to have the lower body of the Incredible Hulk and the upper body of Kate Moss?" he asked.

I ran the theory by Polly.

"We want you to be strong enough so that it feels like you're jumping off a solid concrete platform," Polly told me. "All your energy should be going upward efficiently. If you're not strong in the core as well as the legs, if you're leaking out energy all over the place"—and here she kind of waggled her body—"it's like you're jumping out of a rowboat."

We were joined by Jamie, a lightly tanned, sunny, small, and muscular trainer who looked like a shiny young soccer player. "We're going to get you to dunk in a year, right?" he said with an enthusiastic laugh. "We're putting together a fifty-two-week regimen for you." He had gotten up at 5:30 that morning to get to work especially early to finish it up. Holy Jesus. I get up that early only if I have to catch an airplane. As I lay myself down to do some crunches—the first test of my core strength—I suddenly didn't want to disappoint these two people. They already believed in me and in the project more than I did. Beneath those fluorescent lights, I told myself I would succeed—I would slam the ball home.

Then came reality: I could pull off only a measly half-dozen sit-ups before my hips started to lift up and my feet came off the floor. I whinnied, not unlike a horse. Polly checked the "Needs Improvement" box on her clipboard.

She measured my grip strength. The metal vise gave disconcertingly little ground as I squeezed. I felt like one of the guys who had given it a shot before Arthur came along and pulled the sword from the stone.

Next, Jamie and Polly demanded what amounted to an impossible push-up. Dear reader, you should try it: Lie down, face on the ground, put your palms facedown, with your thumb-tips by your eyebrows, less than an inch from your forehead—now complete a push-up. Ain't happening, right? There I was, lying on the ground, trying to figure out a way to push myself up. Anything to save face in front of my two trainers. I got, perhaps, the slightest insight into paralysis: intellectually knowing what must be done but not able to even twitch. I let out a little grunt. "Oh, that's totally OK," Polly said. "Yeah, most people aren't able to do even one," Jamie chimed in. "Just try some regular push-ups."

They announced they would next measure my standing vertical—a jump test—to get an early sense of my jumping ability. Jamie positioned me with my left shoulder and side next to a red cinder-block wall and asked me to reach my left hand straight up; he marked the spot with a bit of masking tape. Ninety-eight inches high with my feet planted on the ground. He then wrapped a bit of tape around my left forefinger and middle finger. I jumped, slapping the tape against the wall at the highest possible point. The distance between the two pieces of tape: 18 inches, less than half of Michael Jordan's vertical at his prime. While the standing two-footed vertical isn't an exact substitute for a running leap—which nets me another four or five inches (I could just touch the rim, 120 inches off the ground)—it'd be a perfect proxy for the progress of my explosiveness. Given the size of a basketball, I'd have to get my hand at least 126 inches off the ground to slam the ball. In short, if I could increase my standing vertical by five inches, or my running leap by eight inches, about the height of this book you're reading, I could dunk.

The two then led me through seven exercises that would assess my mobility and stability. The first was the overhead deep squat: Holding a dowel overhead, elbows locked, I began lowering my butt to the ground. I couldn't even get my thighs parallel to the floor. "That's all right," said Polly, as I pathetically tried to get lower. "We'll just have to work on your flexibility." That moment, standing bare-chested and fluffy-haired in my shorts and sneakers, I felt a pang of embarrassment. The In-Line Lunge, the Hurdle Step, the Active Straight Leg Raise. None went beautifully. All afternoon I kept hearing things like: "The support leg is slightly wobbly on both sides." Or: "Your left knee tends to collapse inward and your hips drop out of alignment."

It sounded like I was an old man. Misaligned, in addition to being sub-functional. Maybe I shouldn't have been surprised. I think of myself as a lumberer. "You walk as if you think the floor is one foot deeper than it is," Rebecca tells me.

The best possible score on the Functional Movement Screen, as the set of exercises is known, is 21, though that's unusual. "Increased risk of injury may be associated with scores < 14," said the evaluation material Polly later sent me. Then: "Your final score is 11."

"It looks like I'm jumping out of a rowboat," I told her.

She corrected me, with a laugh: "A leaky rowboat."

•

After we wrapped up tests at the Performance Lab, Polly told me we would head to a place called the Motion Analysis Lab. "Everyone over there looks like they won the science fair," she said over her shoulder as she spirited me from one building to another. "I always feel like one of the monkeys." Earlier that very day, she told me, some guy was hooked up with sensors as he practiced, 100 or so times, wristing a dart toward a target. The motion was meant to test recovery from broken wrists, she explained, as if most people snap their wrists like they're barflies.

Polly had shuttled me to the lab to do something far more prosaic—and painful: a Wingate Test, which is meant to test peak power in anaerobic conditions. I would pedal atop a stationary bike for five minutes, with little resistance. Then, thirty seconds of biking through three-foot-deep mud—and I'd have to pour myself into slogging ahead as fast as I could. "Pretend you're dragging an elephant," peeped a slight, bespectacled graduate student. As if I'd dragged an elephant just the other week. As if dragging an elephant were humanly possible. As I got on the bike, the student discreetly pulled over a small wastebasket. "Just in case," she said.

The namesake of the Wingate Test, Orde Wingate, was a slight British soldier who won fame for his derring-do in the 1920s and 1930s. I think of him as a cross between Lawrence of Arabia and Rambo. He had the self-confidence in battle to give himself the radio transmitter call sign "Winner." He was a renegade, a wisp of a soldier, a Christian zealot, a hard-ass, and something of a romantic.

Above all, he was "fanatical about fitness," according to his biographers. A lifelong friend reported that as a teen Wingate "had no use for girls whatsoever and concentrated on physical fitness and his horses." Wingate himself once described his excitement about going horseback riding this way:

> "Today," I said to my soul, you will forget this tawdry life of reality, this vulgar present, this banal undistinguished existence . . . such trivialities as "the great heart of the people," "liberty, fraternity and equality," "humanity," "evangelicism," "the dictatorship of the proletariat," and other dreary coarsenesses! Today, in the only world I will admit

21

to be real, the real world of dreams, of fancies, of departed glories brought nigh, in this world ... you are about to move. Black is the past, black may be the future, but the present, oh my courageous soul, is glorious. Today I shall be like a god."

Ah, yes, just how I feel each morning when I throw a towel around my waist to grab the morning newspaper. Godlike always, that's my motto.

Serving in India, the Middle East, and Burma, Wingate became known for his dashing raids in the colonial spirit, and his feats of endurance. A 1933 journey through Libya in search of Zerzuza, a mythical oasis, "was desperately hard going, made even more so by the extra hardships that Wingate had devised, true to his habit of testing himself and those under him to their limits," write John Bierman and Colin Smith in *Fire in the Night*, about Wingate's exploits. "Spurning the usual routine of desert travel, he made his protesting men and complaining camels plough ahead through the heat of the day, instead of traveling through the night." I'm sure his men adored him. In another epic test, he led his men through fifty miles of thick Burmese jungle in 48 hours. "The inexperienced soldier," Wingate wrote, "is apt to think that where he suffered there is something wrong instead of realizing that suffering is a normal and necessary part of operations."

Wingate died in a plane crash in jungle hills in India at age 41. Lord Mountbatten, charged with overseeing Wingate's operations, as if any mere human could do so, eulogized him as a "fire-eater." Peggy Jelley, with whom he had a five-year affair in his late twenties, never married because "after Orde, all other men seemed uninteresting." He had done some of his most fearless fighting in the Middle East, and after the creation of the state of Israel the country's newly minted sporting and athletics center, near the Mediterranean city of

Netanya, was named for him. Scientists at the center developed the Wingate Test, meant to evaluate the very qualities that made Orde Wingate such a badass—power and endurance.

Judging by the test named for him, I might have mutinied had I been one of Wingate's soldiers. I didn't vomit, but my legs afterward felt like only recently refrigerated pudding. My head hurt and my chest heaved. My results on the Wingate Test, as with everything else I did that day, were mediocre. For good or bad, I was a decidedly everyday athlete. My mean power was 607 watts; during the slog-through-the-mud part, which was meant to simulate the sort of explosion of power needed to do things like dunk, it reached 770 watts. The lab scientists offered some bizarre comparisons to explain my performance, apparently the only ones they had on hand: I was in the "very good" category of Israeli males my age, and my results "were consistent with those of female Division I basketball centers." OK, there are recorded cases of women and Jews (like me) dunking. But not very many. (My colleague at the *Austin American-Statesman* Ralph Haurwitz tells a story of the week his daughter got into Harvard and his 15-year-old son first dunked. Another of our colleagues, also Jewish, joked: "What's the bigger headline? 'Jewish Girl Gets into Harvard' or 'Jewish Boy Can Dunk'?") Part of me was a little disappointed. I was never a lousy athlete and I had secretly hoped that I would turn out, in this initial reckoning, to have hidden talent. But empirically, I had just learned, I was completely average. Clearly, if I could dunk at the end of the year, my achievement would be a testament to the malleability of the human body.

Before I left the Performance Lab that early summer afternoon, I suggested Polly and Jamie privately scribble down and seal their thoughts about whether I could ever dunk. Then, summer-camp-style, I'd open up the envelopes after a year had passed. I didn't want them to reveal to me their doubts, but I wanted, eventually, a frank sense of their uncertainties. Polly waved my idea off. "I really think you can do this. I see people do things they think are impossible all the time here." She then told me a story. She once asked an old coach of hers whether she ought to compete in a triathlon intimidatingly called Escape from Alcatraz. If she couldn't make it across San Francisco Bay, she didn't want to bother. "Whether you think you can, or you think you can't, you're probably right," her former coach told her.

That night Rebecca and I and a friend bought $15 last-minute tickets to the New York City Ballet. Sunk into the plush seats, my arms sore from my pathetic dozen push-ups, my legs as wrung-out as an old sponge, I found my weary eyes tracking the male dancers as closely as they could. Stage lights twinkled behind them. Peering out from the dark, I watched the clenching of their buttocks, their pinpoint balance, their controlled, precise jumps. Telescoped before me were decades of training, of dieting, of stretching. This, I thought to myself, would be a hell of a year.

2

Evolution and the Dunk

During his senior year with the Illinois College Blueboys, crew-cutted guard Jacob Tucker averaged more than 14 points and seven rebounds per game. He earned captaincy of the team and the respect of opponents. But it wasn't for his basketball skills that the obscure Division III player became an Internet phenom. In March 2011, the 5'11" Tucker put video of himself dunking over friends on YouTube. His jump was measured at an eye-popping 50 inches. The video went on to capture five million hits, and the NCAA invited Tucker to compete in its Final Four weekend slam-dunk contest. He won—and became a small-time celebrity. After throwing out ceremonial first pitches for the St. Louis Cardinals and Chicago White Sox, he was signed by the Harlem Globetrotters. Under the name "Hops," he now runs the pick-and-roll a thousand times a year to "Sweet Georgia Brown" with a 7'8" dude named Tiny.

There are many ways to determine the greatest jumpers. There are the famous dunkers, like Michael Jordan and Dr. J, who won the adoration of millions on television. There are the inch-for-inch leapers, like wee Spud Webb. There are the artists, like Mikhail Baryshnikov and Anna Pavlova, the original dying swan, whose successes are measured in gasps, bravos, and applause. Then there are the Olympic-caliber athletes, like the Cuban Javier Sotomayor, perhaps the prettiest of jumpers, whose 1993 high jump world record of eight feet and one-half inch, set with a series of long, languorous strides in short shorts, still stands. Or Bob Beamon, whose 1968 Olympic long jump of 29 feet, two and a half inches, in the high altitude of Mexico City, took 15 minutes to measure because the optical device on hand wasn't equipped to measure such a feat, and who, when he finally heard the news, melted to his knees in shock.

But for all the tremendous feats of these athletes, humans generally have terrible explosive powers, whether jumping or sprinting. "Humans are mediocre runners in several respects," Dennis Bramble and Daniel Lieberman wrote in an influential 2004 *Nature* magazine article titled "Endurance Running and the Evolution of *Homo*." "Even elite human sprinters are comparatively slow," they continued, capable of sustaining maximum speeds of 10.4 meters per second for no more than 30 seconds. In contrast, other mammals, such as horses and pronghorn antelopes, can gallop 20 meters per second for several minutes. Even at short distances, we're pretty inefficient sprinters compared to our mammalian brethren: In top gear, our bodies demand roughly double the metabolic energy of similar-sized mammals. Compounding matters, human runners are less maneuverable and lack the many "structural modifications" of other mammals, such as feet in which the toes are on the ground and the heels are off (as is the case with dogs and cats) and short limb segments, that make them good sprinters and jumpers.

But what humans are good at—fabulous, really—is endurance running. That quality is unique to humans among the primates. It's uncommon among four-legged mammals, too: Well-conditioned human runners have even been known to outrun horses over long distances. Anyone who is a dog owner knows that his or her dog is much faster in a short sprint. Even fat dogs. But take them for a long run on a leash on a hot day in the park, and they'll eventually refuse to move. Panting, they'll splat out like little polar-bear rugs on the sidewalk.

Stride length, skeletal strength, trunk and head stabilization, heat regulation—even reduced body hair compared to our forebears— make us distance-running machines. With each step, our collagen-rich tendons and ligaments store elastic energy during the "braking phase" and then release it through recoil during the ensuing "propulsive phase." Your leg basically acts like a spring, coiling up and then releasing out. "We actually jump all the time," Lieberman told me when I reached him at his Harvard lab. "It's called running, which, in effect, is a series of very small jumps."

But as evolution selected us for endurance running after we clambered out of the African trees and onto the plains, it de-emphasized explosive jumping ability. All animals—whole species as well as individuals—face this trade-off between slow-twitch muscles and fast-twitch muscles. You can't be both a great long-distance runner and a great sprinter. And very long ago, *Homo sapiens*, or, more likely, our predecessor *Homo erectus*, made that swap. Our ancestors, chimps, remained arboreal, climbing from tree to tree and branch to branch, and studies show that other primates have much better jumping ability than we do. A 2006 paper by a group of Belgian scientists, for instance, found that bonobos have verticals nearly twice that of humans.

Yet this loss of jumping ability and the evolution of endurance running was crucial for human development. Before the advent 30,000 years ago of sophisticated projectile technology, such as bows and arrows and spears, that would allow us to hunt from afar, long-distance running ability meant that we could chase our prey to exhaustion. Endurance running, write Lieberman and Bramble, may have "played a role in helping hominids exploit protein-rich resources such as meat, marrow and brain first evident in the archaeological record at approximately 2.6 million years ago, coincident with the first appearance of *Homo*." ("Marathon runners may be a masochistic lot," the Duke biologist Steven Vogel once observed, "but they don't do anything physiologically perverse or historically novel.")

In a sense, then, my dunking quest meant not only changing that balance on an individual basis but fighting against some of the very things that evolutionarily made me human. Which was exactly why, at the end of our interview, Dan Lieberman laughed and said that I was destined to fail.

•

In June, my New York City journalism fellowship over, I made the long drive, with wife and dog, back to Texas. Some people call Austin, my chosen hometown, the Velvet Rut—a place where people settle a little too easily into their comforts. Maybe it's the hot, languid summers. The land of the lotus-eaters, my brother-in-law calls it, recalling the island whose distracting flowers steer Odysseus' men toward oblivion as they forget their homeward purpose. Even the smoky-smelling meat takes its time getting cooked: In Texas, barbecue means exposing brisket and ribs to indirect heat for fourteen hours or more as they slowly transform into what Rebecca calls meat candy. This is her native seat—Rebecca's parents have long taught at the university—and, once we wrapped up college and grad school, she wanted to come back.

Rebecca had moved around a lot as a kid, with her parents conducting research or doing visiting lectureships abroad or in California, but the family home in Austin, an oak-floored 1920s place with a backyard shaded by pecan trees, was always the touchstone. The moving around made the family very close, and Rebecca, the littlest and youngest of five kids, a little shy, always thought of herself as the runt. She's not actually that small—she's 5'6", maybe 5'7"—and in private she's something of a Muppet, pulling silly faces and impersonating Crocodile Dundee, but she still scrunches up easily, finding ways to curl up beside me even on a Greyhound bus. She's the sort of person you could fit snugly into a pantry cabinet, if you ever had to put a person somewhere up high. And honestly, she really wouldn't mind; she's hardy and resourceful, as handy with knitting needles and a hammer as she is with a pen. If you prepared her a little bed and left her a book, she'd be satisfied for a long while.

We had met at Yale early in our freshman year—her twin sister and I shared a philosophy class; soon the three of us had a steady lunch date after the seminar—and by Halloween I was smitten enough that I allowed her to dress me up as a die-hard Dallas Cowboys fan. (Rebecca still spends her Saturdays and Sundays each autumn wearing Longhorns T-shirts and Cowboys jerseys, ever ready to chew out boneheaded, head-setted offensive coordinators on television.) I'm pretty sure that as a New York City kid I didn't know Austin existed before I met Rebecca. If I knew of it at all, it was as a piece of trivia—one of those random cities you wouldn't expect to be capitals: Frankfort, Kentucky; Harrisburg, Pennsylvania; Carson City, Nevada. When I first got here, just out of school, I thought I would stick around for only a year or two before seeking my fortune back east. I was hungry. Soon, I was seldom hungry, partly because I ate so much and did so little: the television, the Tex-Mex, the laptop, another juicy peach, a spoonful more of whipped cream, ice cream, too, another helping of barbecue. Austin grew on me. "I haven't got one of those bellies that sag half-way down to the knees," George Bowling, George Orwell's narrator in his midlife crisis novel *Coming Up for Air*, tells us. "It's merely that I'm a little bit broad in the beam, with a tendency to be barrel shaped." I now had a chance to rediscover some get-up-and-go. But how much youthfulness did I have left in me? Besides the problem of evolutionary fate, I faced a more immediate challenge—my age. At the still-tender age of 33, was it too late for me to dunk?

"You are definitely on the cusp," Susan Brooks, a professor of molecular and integrative physiology at the University of Michigan, told me. She specializes in muscle function—and failure—among older people. "When we talk in generalities—and there are exceptions to every rule—age 30 is right around peak in performance, in terms of body mass and strength. You are not by any means hopeless—bad stuff with muscle tissue doesn't really happen till age 50."

"But age 30," she repeated, maybe just to depress me a little, "is when we start to see declines" in how efficiently cells process energy.

Each of us is born with a set number of muscle fibers, and those can change in length as we grow, or in diameter as we train. Or the fibers can shrink. They're very plastic. You can modulate the size of those individual fibers throughout your whole life, especially through exercise. But around my age, and certainly by 50, some of those fibers will start to go away, in a process that isn't very well understood by scientists. By the age of 85 you may have lost as many as half of your weight-bearing muscle fibers, like blocks pulled out of a Jenga tower. Typically, the fibers that get lost early are the fast-twitch type, the ones for explosive power. That's why sprinters typically fade at an earlier age than marathoners. And why the average age of NBA players hovers around 27. Researchers have noticed this phenomenon even in some of the longest-living creatures on Earth— bivalve mollusks—and as different as they are from humans, they still tell us something about ourselves. The clamping closed or the fast release of a shell is not, in its way, unlike the explosive power needed to dunk a basketball.

Tendon elasticity—an important component of jumping—also narrows with age, as the quality of collagen changes. As we prepare to jump, in that second in which we lower our butts and push out our knees, we stretch our Achilles tendons by about 5 percent. That tendon stretch stores up the energy like a stretched rubber band before it's released into the air. But we're far less flexible as adults, of course, than we are as babies. And in our 30s we're less flexible than we were as 20-somethings, meaning, ultimately, that we have less capacity to hoard energy before embarking on an activity like jumping. The early 30s is when the nerve impulses begin to quiet down: By the age of 55, my nerves will fire 5 percent less quickly than they do now. All of us, in short, are turning into earthbound statues by the day.

Still, the plaster is not quite set, and researchers have recently found that things thought to be beyond improvement can, in fact, change with training. The operative word here is "train." And if I was going to dunk, my charge was to train like a madman. Jamie had just sent me an Excel spreadsheet with a regimen. "We're so excited to be on this journey with you!" he wrote me. How do these guys always make pain sound so metaphysical?

•

The sanctum sanctorum of the University of Texas basketball program is the basement of the Denton Cooley Pavilion, a brutalist building adjacent to the Drum, the school's basketball arena. I had scootered up on the sort of super-hot Austin day that reduces a person to the consistency of baking cookie dough. A glass cabinet held player-of-the-year trophies belonging to Kevin Durant, the NBA star and former Longhorn, and conference-championship bling belonging to the rest of the team. "Desire Creates Power" read a sign inside one of the cases.

I was there to see Todd Wright, the program's strength and conditioning coach. The NBA player and former Longhorn LaMarcus Aldridge once said of Wright, "He had a plan that wasn't like any other weight training. He looks at feet. Feet. Hips. Everything. It shows that he knows how the human body works." An *Austin American-Statesman* article described his methods "to help a player learn to stretch parts of his body so complicated actions—the proper release of joints, the firing of muscles, the preservation of energy—are as efficient as possible." If I was going to dunk, Wright might be my Yoda.

"He's part of my family," Durant once said. In a sign of how strongly his players and bosses believe in Wright's work, he will take home $235,000 in pay this year. (The head men's basketball coach, Rick Barnes, is among the state's highest paid employees, with an annual salary of $2.4 million. This being Texas, university football coach Charlie Strong is the highest paid, at $5 million per year.) I wanted Wright to help me, but first I had to get to his office, in the weight room at the base of Denton Cooley. Beyond the glass cases I found myself facing a locked glass door, one that required a special key code to get through. A clean-shaven young man wearing a burnt-orange polo shirt punched in some numbers and in I went. Perhaps UT's weightlifting facility is the world's safest, I thought, the Fort Knox of worthless metal.

Wright was intimidating, a bison of a man. Six-foot-four and brawny. His head was about the size of a small watermelon, and just as bald. Clear blue eyes stared out from heavy folds of skin. He told me to sit down. Like a dog, I shrank into a chair. To dunk, he said, you need to work on three things: ground reaction, overcoming gravity, and momentum. "These are the drivers," he said gravely. I asked him what he meant by "ground reaction." "We want you to push your feet through the ground as fast as possible," he said. "We want your arches exploding." That sounded painful. He had put together a theory of jumping improvement that involved jumping, hopping, leaping, and skipping, which he had boiled down to the ungainly contraction "juhoplepsking."

He looked me over. "You definitely have the architecture of a dunker," he said approvingly. I admit to a swell of pride—as if I were any more responsible for my height than for my name. We began talking basketball, and he naturally started mimicking players' motions from his swivel chair. I was surprised at his grace. He demonstrated Dirk Nowitzki's shot, leaning back in his chair, kicking out one leg, and then flying a very lanky arm straight back, almost behind him, and snapping his wrist through the air as he shot an imaginary ball. He looked like he was swimming backstroke.

At one point, he fished in a sack for something, and then pulled it out: a skeletal model of the lower leg and foot. And this large man, a former college middle linebacker with a deep, baritone Massachusetts voice, began gently manipulating the foot as he held forth on the nuances of muscle and foot coordination.

●

A few weeks later, on a Thursday evening, as the thermometer dropped just below 100, I steered my car west, out of Austin, into the rolling scrub-brush landscape known as the Hill Country. I passed ranches with old stone walls and barbed wire, over creeks dry in the summer drought, and past signs that read "Burn Ban in Effect." The governor had recently asked everyone to beg God for rain. I was looking for the Fitzhugh Baptist Church, one of dozens of churches that have sprung up in this corner of Texas to serve an increasingly suburban population. The churches often have catchy signs to attract passersby, and Fitzhugh, cut out of a stand of live oaks, was no different: "Hot? Church is Prayer-Conditioned." I was there not to pray for rain but to watch some basketball.

Inside the gym beside the church, Chris Corbett, a tall, pot-bellied, salt-and-pepper-goateed 40-something private basketball coach was putting a half-dozen kids through a series of curl drills. Corbett had kindly invited me out here after I contacted him, asking him to put me in touch with kids on the verge of dunking. (I had decided to go through a coach because the alternative—lurking at community courts to talk to young, promising teenagers—could come off as pervy.) He wore knee-high black socks and a knee brace, and even though the gym was thoroughly air-conditioned, his shirt showed the faintest bit of moisture as it clung to him.

The kids were all young teens trying to get onto their high school squads. Silently, obediently, they executed the drills, or tried to, as Corbett shouted instructions and encouragement. Their parents sat on some folding chairs on one side of the court, making small talk with one another, staring out into space, watching their kids, or reading. One kid, Cabe, about 5'7" with a thatch of black curls and a sleepy-looking face, was absently palming a ball between drills. I told his mother, Lauri, that I wished I could do that. She said the boy was already wearing size-13 shoes. "I'm just hoping he grows into that body," she said. Cabe had some talent, but he seemed indifferent to the coach's orders. Finally, as he made his way around a chair too slowly in a drill that simulated a pick-and-pop cut, Corbett lashed into him: "You taking a break to have a margarita?" The best players were self-assured sisters, aged 15 and 13, Alé and Mariana. In her camouflage shirt, Alé drained a series of jumpers over a tall, thin boy who played a little shyly. ("When she got here two years ago, she could steal but not score a basket for all the gold in China," Corbett told me.) Mariana, a fierce girl only a shade over five feet, kept feeding the ball to her older sister for open shots, giving small pumps of her fist each time Alé made one. Their mother whispered to me that each night before Mariana gets into bed she prays to God for height.

I was there to see two kids, each on the brink of dunking. Josh Scoggins, a 5'11" 15-year-old, was trying to get off the bench of his suburban high school team. As he warmed up he told me that his favorite dunker was Michael Jordan. Hadn't Jordan retired before Josh was old enough to really watch basketball? He had seen videos on ESPN, he said, grinning. Josh had a thin, sinewy body, one in which he seemed very comfortable, and he began a series of take-offs in his new Air Jordans. He was trying to dunk a volleyball, and he was successful maybe three times out of a dozen. His mother, a bright-eyed woman, began talking my ear off as I scribbled notes. Had I heard of a particular book on performance management? she asked me. She asked me about my book—what was the *value* of it? she kept asking. She asked what I wrote about for the local newspaper. Energy and the environment, I told her. "Green everything is a hoax," she harrumphed. "My motto is as few rules as possible and as many as necessary. That way we stay the freest nation on earth."

I fled to the other end of the court, where a smiley-faced boy named Laquan Williams was trying his hardest to be assertive in the paint. Another coach, a former UT player, was walking him through post-up drills. Laquan was 14 and a linebacker at Hendrickson High in Pflugerville, a neighborhood north of central Austin. He told me, through deep heaves, that his coach had told him to pick up basketball to learn some footwork. I could see why: he moved like a baby giraffe, an awkward procession of limbs. Just over six feet tall, he wore a beat-up pair of Nike running shoes. He was strong-looking—big arms that made his T-shirt look tight—and narrow-waisted. I teased him that he was dogging his drills. He hadn't gotten much sleep last night, he sighed, hooked on one episode after another of a Nickelodeon cartoon. I talked to his mother, a former volleyball and basketball player in her early thirties who now served in the National Guard. "He's a man physically, but he's growing mentally," she laughed as she handled her grandson, nine-month-old, mohawked Carmelo Caesar Anthony. (He's Laquan's 17-year-old sister's kid.) Laquan grew up without a father, and his grandparents recently died. "He thinks he needs to be the man of the house! He tries to seem intense, but he's really not."

When Emerson wrote that "America is the country of young men," he had in mind the can-do buoyancy of would-be dunkers like Laquan and Josh. (Not that Emerson could have known what a dunk was, of course.) And as I stuffed my notebook in my back pocket, I wondered whether I could lasso their optimism and drag it back to Austin with me.

•

I myself was never a born jumper—that's the idiomatic way of putting it, as if it's our fate, written in the stars at birth, never to dunk, literally and metaphorically. As if, in the end, some things are simply closed off to us. Or are they? I've had reason in my life to ponder this question. When I was at the peak of my fitness and youthful promise, about the age that Spud Webb was as he flew hoopward and dunked himself into my consciousness, fate introduced me rather abruptly to my own physical limits, and science introduced me to the ways we humans sometimes can outwit them.

One sunny, clear morning in early 2006, six years before that afternoon in Central Park when Nathaniel and I batted around my dunking odds, I found myself on a hospital room gurney, a sonogram technician wincing at the ultrasound image before him as he passed his wand back and forth over my gelled left testicle. Whatever my potential might have been, it was now naturally circum-scribed, I would soon learn, by an unforeseen cellular mutation that would, if left to its own devices, insist on multiplying, voraciously, until it killed me.

I didn't know it at the time, but the first sign of my testicular cancer had come a month earlier, an inexplicable backache I woke up with the morning I was supposed to play tennis with my buddy Alf. I was 25. I felt sheepish, a young man canceling a tennis game because my back hurt.

"You're not going to believe this," I remember chuckling on the phone, "but my back is killing me."

As mysteriously as the backache surfaced, it subsided. I barely had a chance to think about it anyway. Rebecca and I were negotiating to buy our first house—a fixer-upper in Austin. We had made the mistake, while still renting, of curiously popping by open houses: to snoop, in a sanctioned way, around intimate spaces; to be privately snobby about other people's décor; to envy, finally, what they had. It's a dangerous game. Soon you find yourself with a Realtor, a home inspector, and a contract.

Then I noticed the lump. I was soaping myself in the shower and found a hard bit, the size of a half-pea maybe, bulging from my left testicle. I wondered if it had always been there, like some hard-to-see birthmark. I decided to ignore it.

A couple of weeks later, on that sleepy Saturday morning in January, I woke up again with another strange ache—this time in the left side of my lower abdomen. It was a dull, quiet, thudding pain, as if someone had kicked me in the groin.

I hadn't told Rebecca about the lump. My stomach was hurting, was all I said as I shuffled off to the bathroom to see if the lump was still there. It was. It didn't seem any larger, but now I was worried. Maybe, I called from the bathroom, we should go to the hospital.

"For a stomachache?" she asked.

"I don't feel so well," was all I managed, suddenly unsure of my own body.

After the sonogram technician completed his inspection, I was wheeled back into a hospital room, where the on-call doctor told me that I had testicular cancer. "Exactly what Lance Armstrong had," he said, as if that would make me feel better. He wore a Daffy Duck tie. "So some people get this and live."

As Rebecca started crying on my chest, the certainties of the world suddenly uncertain, I giggled and wept. I giggled because the whole situation seemed absurd; I wept because I knew my parents would be breathtakingly sad, and I couldn't bear the thought of telling them. My dad had been diagnosed with prostate cancer a month earlier; this was something else they didn't need.

The chief ingredient in the testicular cancer chemo cocktail, I learned, is cisplatin, essentially liquid platinum. (The story goes that a biophysicist, experimenting with bacteria, unexpectedly found their growth stymied near platinum-plated electrodes he was using to stimulate them with electricity. That discovery would be taken up by a cancer scientist who began developing platinum-based drugs to inject into humans.) Slouched in a La-Z-Boy, diminished and bald, the backs of my hands and my forearms sore from the daily needle jabs, cisplatin coursing through my veins, I used to think of myself as one of those pathetically weak comic book characters secretly gaining superpowers—Platinum Man, *impenetrable and unbreakable*. A fantasy, of course, but, after all, the drugs were allowing me to sidestep the misfortune of my nature. These were not exactly Spud Webb–like superpowers (*leaps to the basket in a single bound!*), but they were superpowers nonetheless.

I survived. After my testicle was maneuvered out through a slit in my abdomen by a urologist named, I kid you not, Dr. Dick Chopp; after three cycles of chemotherapy; and after all my hair fell out, only to grow back in unexpected thick curls so baby-soft as to announce a second life, I was declared cancer-free.

The Greeks had conceived of the Fates as severe women, crooked with age. Three nights after a child's birth they decided the course of his or her life, spinning, measuring, and cutting the thread of life. In an earlier time, without the intervention of medical science, my fate would have been an early death—a short thread. Instead, in a surprisingly short period of time, I was fine. How much, I wondered, of our story do we humans weave with our own hands? How much can we manipulate the fabric of our physical lives?

PART II

THE YEAR
of the
DUNK

3

The Dunking Year Begins

Aug. 25, Eve of the Year of the Dunk: *No. of tortilla chips consumed at Polvo's, at roughly one every six seconds for nine and a half minutes: 95. Enchiladas de la casa, plump with gooey white queso: Tres. Plus rice and refried beans. Followed up with a root beer float at Amy's Ice Cream. Weight: 203 lbs. Hasta luego, dessert. 365 days to go.*

Two days into my grand test of human capability, on a muggy midweek evening in late August, it seemed obvious to me that my capabilities wouldn't budge. I felt like someone had unfairly reorganized my musculature. As if my biceps, my shoulders, my hamstrings had been jumbled up and then clumsily knotted back together. Muscles that I hadn't known existed ached in rebuke. I wasn't all that surprised. Those first couple of days I did four hours of weightlifting and squatting. The dumbbells might have been about the size my septuagenarian parents used, but they definitely qualified as weights. So unused was I to pumping iron, the very bones of my hands throbbed after the first set of repetitions.

Carving out the time was not easy. I had to slip away from work.

Get up early. Go in on a Saturday. Getting myself into dunking form wasn't shaping up as much fun. It was certainly a narcissistic exercise. I caught myself, half-naked, staring into mirrors, squeezing my sides to figure out if they had tightened up any. I examined my belly for any nascent signs of a six-pack. I turned profile to see if I was looking any slimmer.

I did my daily work at a gym set up within the newspaper where I worked. It was called the Press Room. The brass had installed it a couple of years earlier as part of a wider effort to make employees feel better about themselves. Me, it made feel weak and pathetic. But I had to give the publisher credit—it was an impressive gym, with lots of shiny equipment of the kind that I had studiously avoided the previous 33 years of my life. It also offered a small program of classes that I was determined to attend to increase my flexibility: yoga, Zumba, cardio club.

That first week I made sure to stop by yoga. Six women with their mats, led by Kandice. I stepped into their dimly lit room a couple of minutes late, just as Kandice was leading them in Horserider's Pose, a kind of lunge. I grabbed a mat and tried not to distract as I stomped about as lightly as I could manage.

Kandice's head spun around: "Your sneakers," she hissed.

"What about them?" I whispered.

"This is yoga," she scolded. "We don't wear sneakers."

Well, then. I took my sneaks off as quietly as I could and struggled through simple postures like Downward Dog. A small redheaded woman was positively canine in her execution. I comforted myself with the following thought: No way can she dunk.

"Feel your internal organs," Kandice told the group.

The only organ I like to feel is my external one. Otherwise, I prefer to think of my innards as a single blob, if I think of them at all.

"Move into Child's Pose," she announced. In this exercise, you kneel on the ground, drop your butt toward your heels, and stretch your head and hands forward. Except my butt didn't really drop. It just stayed up like a camel's hump. Kandice stepped off her pad to push my butt toward the ground. It didn't want to go. She shoved hard, and under her touch it stiffly moved toward my heels. She moved away, and, like a balloon just let go by a child, my bum began steadily rising skyward on its own.

●

I would need help cutting fat, too. I was still hovering over 200 pounds. Not bad for a guy my height, but not good if you want to dunk a basketball. Like domesticated animals, we have grown fattened and slow, far removed from our ancestral, lean hunting selves. Steve Austad, an expert in the biology of aging, tells me there's a marked difference between the athleticism of wild mice and lab mice. "Lab mice are lucky if they can jump two inches," he says. "Wild mice can jump out of a garbage can. There's something in that domestication that leads to the loss of explosive abilities."

I want to be able to jump out of the garbage can.

To cut weight, I turned to my friend Phil, whom Rebecca affectionately calls the Incredible Shrinking Man. We went to high school together, where Phil, in his good-natured way, struck a character both jovial and cutting. He was witty and quick and, famously in our circle of bleeding-heart liberals, an unabashed Reaganite. The man actually had a framed photo of the Gipper in his Westchester bedroom. He was unathletic and a little overweight, but he was a nasty Ping-Pong player. He played like a stick figure, getting the lines of his body just right, if a little rigid, to put away most shots. His parents kept a table in the basement of their home, and I blamed my never-ending losing streak on this fact. (I similarly put my mediocre basketball skills down to my not having a hoop in my driveway. Not that we had a driveway, for that matter.) Once I almost beat him. I had a lead late, and was about to serve, when Phil told me that a girl I'd obsessed about for years was interested in me. My game was shattered. Phil would become a white-collar criminal defense lawyer, using his fine theatrical instincts and formidable will-to-win to represent insider traders, wayward military contractors, and big-ticket fraudsters. If you read about them in the newspaper and shook your head about the nogoodniks of this world, he or his firm may well have represented them. (*Alleged* nogoodniks, Phil would winkingly say.)

He also carried around with him that extra weight. He didn't mind it, as far as I knew, and, of course, no one else did, either. But a few years ago, a few of his officemates decided to put together a betting pool on who could lose the most weight, as a percentage of total body weight, in six months. The competition of it aroused Phil. He would, he quickly decided, obliterate them. "Instead of the gong for dinner, let us hear a whistle from the Spartan fife," Emerson wrote in *Self-Reliance*. A shrill piercing noise evidently rang in Phil's ears those six months: He shed sixty pounds—more, really, than he had to lose—with a diet of granola and nonfat yogurt in the mornings; a salad with light oil and balsamic vinegar, tossed with avocados, for lunch; and a dinner of lean chicken sausage with grilled peppers and onions—no oil added, of course. "It was really boring," he says. "But I wanted to win."

His advice to me was straightforward: Cut the carbs, and work out an hour each day—*like it's your job.* For me, working out had always been sport; sweating was something I happened to do while playing with a soccer ball, a Frisbee, or a basketball. I would have to endure smaller, healthier portions and a steady diet of zero-fat yogurt and nonfat milk—the devil's water, I call it. Now, it was becoming clear to me, what had begun with the feel of a lark was growing occupational.

•

And so it was, that fall, on the lookout for a spot outfitted with some of the equipment lacking in the Press Room, that I found myself stepping for the first time inside a commercial gym, one wedged, like so many things in Austin, into a strip mall, between a Taco Bell and an X-ray clinic. It was a miserable place. Humans, in all their wonderful messiness—rolls of fat, knock-knees, wonky postures— were groaning to transform themselves into efficient robots. Here was a bank of treadmillers, running like a row of pistons, *ker-plunk, ker-plunk, ker-plunk*. There, a man was seated on a black, sweaty, faux-leather stool, swiveling unhappily away on an abs machine. Heavy bass throbbed from overhead. Most people were expressionless, drained of their humanity. Maybe they were thinking about the memo they had to write their boss or the groceries they needed to buy on the way home. Or maybe nothing.

The chief giveaway that these were humans: their stupid T-shirts. "And your point is . . ." was written across the front of one. A palefaced 50-year-old who looked like Lieutenant Data from the *Star Trek* TV show wore a shirt that said "What part of 'shall not be infringed' don't you understand?" One youngish, pimply-looking guy doing triceps dips—as if that wasn't enough, he had tied a chain to his waist from which hung a 25-pound weight—wore a cut-off tee that said, "Nice story, babe. Now go fix me a sandwich."

I tried not to stare. The beefcakes were just so damn beefy. Stonefaced men with stony bodies and bouldered shoulders. Women, too, with enviable flexibility. There was a martial artist who finished her workouts with roundhouse kicks and weird, low-to-the-ground cartwheel, helicopter-like flips. I got a little light-headed just watching.

Going to the gym is a relatively recent phenomenon. In the introduction to his 1889 *Book of College Sports,* famed football coach Walter Camp derided the zeal of "the new Professor Dumbbell, who drags you willy-nilly through a complex system of chest measurement and pull-prescriptions." Three-quarters of a century later, suspicion of the weight room persisted. I once heard Bill Bradley say of his basketball career: "There's one thing we were told never to do: 'Never lift weights. It'll mess with your shot.'" He then deadpanned: "I was thinking maybe I could have had more wins if I had lifted weights." Of course, the notion of self-improvement through weightlifting and gym work eventually won out. It's big business nowadays. Aging baby boomers and the growing number of companies that encourage employees to work out means gyms are now more common than McDonald's.

Terry Todd, a stiff-necked, thick-muscled 70-something former Austin weightlifting champion who had kindly been giving me advice on my training, claimed that an "ideal workout" awaited me each time I set foot in the gym. The trick would be to find that workout, he said. Maybe, I thought, it's like going to a video store. Somewhere in that mess of DVDs there's one that suits your mood perfectly, from which you will derive the most satisfaction. But if you have nothing in mind beforehand, you might find yourself wandering around aimlessly.

I was determined not to become one of the beefcakes striding around the gym. Impressive, yes, but too much. I had a purpose, I told myself—and getting all out of proportion was not part of the plan. Besides, it's difficult to add muscle when you're my size. I mean, that's a lot of territory you're trying to cover. That's what I told myself, at least. And did I really want to add much upper body power? Big forearms are not going to help me get off the ground.

My gym routine opened with a fast jog to loosen up. I ran a 7:30 mile on a slight uphill on the treadmill. Then I stretched, steadily dripping sweat onto my mat. I guiltily wiped it up, then did a set of plyometrics: a series of high-knees, skipping, and jumping. I would do 600 jumps—from a standing position, small, bouncy jumps from my ankles, the sort tennis players sometimes perform before they crouch down to receive a serve; then rapid jumps with only slightly bent knees, in which I tried to get as high as possible; box jumps, in which I jumped two-footed onto and back down from a platform, quickly, quickly, quickly, to develop my explosive ability; and, worst of all, because they really burned up my legs, power jumps from a quarter-squat position, springing into the air and landing back in the quarter-squat. In addition to workouts prescribed by Polly and Jamie, I followed a regimen I bought online called Air Alert that promised its followers a higher vertical. It seemed geared toward bullied teenagers: "Never allow others to interfere with what you strongly believe," read a Zen-inflected booklet that included the regimen, trademarked as Habitual Jump Training. ("Jumping literally becomes effortless like that of a kangaroo.") "Do not allow others to tease or mock your goals or training. Do not allow them to discourage you. Attending to the negative things people may say will convince you to believe that what they are saying is true." The whole thing had a *Karate Kid* kind of feel.

And then the lifting. I did heavy lifting with my legs on Mondays, Wednesdays, and Fridays, giving my muscles time to recover between workouts. ("Let your muscles rest, like fields left fallow," advised Josh, the older of my two brothers and a former marathoner.) I worked my way up the squats, as Tyler, my guru from Central Park, had directed: Each time I showed up at the gym I completed five sets of repetitions at a weight five pounds heavier than I had completed two days earlier. I did single-leg presses, in which, supine, and with much grunting, I pushed a weighted platform away from my chest. I performed so many Bulgarian split squats and Romanian deadlifts that I began imagining that the people of those nations, even the babushka grannies, sported massive thighs and tended the fields in tight-fitting singlets. I did hip exercises, the kind that Polly and Jamie suggested, that had me shuffling from side to side with a giant rubber band tied around my ankles; I did glute bridges, in which I clenched and unclenched my butt cheeks, as if I were dry-humping an invisible creature; I did embarrassing things on an exercise ball, desperately trying to keep my balance as I writhed on top of it, or looking like a hairy, obscene, rhythmic gymnast as, back on the ground, I tried to trade the ball between my feet and my hands. In New York, Jamie had told me that we would "reawaken my glutes." I wasn't sure that was fair to them. All these years, they had slumbered nicely, cuddled up to one another like burger buns. Even asleep they had seemed to do a fine job. Who was I to rouse them? The glutes, along with my quadriceps, abductors, sartorius, semimembranosus, psoas, soleus, and gastrocnemius—among other muscle groups, each more scary-sounding than the last—would be "recruited" for the jump, explained Jamie, as if my body were a Times Square Army office.

Now I seemed to be spinning back the biological clock, growing stronger, slimmer, fitter. Everywhere, I sweated. It was nicely air-conditioned in the gym, but I was like a square foot of rain forest, bearing with me my own zone of tropical precipitation. Undertaking my leaping exercises—up-and-down, up-and-down—I felt like a malfunctioning suburban sprinkler, the sort that pops up irrepressibly and shoots willy-nilly. I was hardly embarrassed; to me the sweat manifested effort, and I couldn't help but feel some satisfaction each time I swept a towel across my ruddy face.

•

At home, I developed a special clothes-destinking system: I draped my freshly sweated-through shirts, socks, and shorts over our front-porch railing; hours later, when they were dry and starchy, I would move them to the hamper in our bathroom; finally, when I had a critical mass of dirty clothes, I would heave them into the washing machine. Despite my precautions, for months our bathroom held the gentlest whiff of boys' locker room. Rebecca, God bless her, said not a word. She was in my corner: However quixotic she thought my mission, she deemed it worthy. A brother of hers, Ben, really could dunk, had even played pro basketball in Germany, and as athletic as I might hope to be, she had seen up close how much stronger and faster and taller he was, and the dedication required to improve. (As a teenager, Ben used to practice into the night on the hoop in their backyard, relentlessly *thump-thump-thump*ing the ball on their little court and rattling the rim; one evening, having had enough, the next-door neighbor took out a shotgun and blasted out the lights. Even dyed-blue Austin is, after all, deep in the heart of Texas.) Her family was pretty much basketball-obsessed, and while Rebecca was always too stumpy-legged to really play, as a teen she briefly kept a diary, one with a Botticelli painting on the cover, documenting Michael Jordan's heroics. In it, she kept track of Jordan's box score, wrote about his dunks, and made rough drawings of the previous night's key moments. She was fanatical about the man—during freshman year of college, a moment in time in which it was mysteriously acceptable and even common to find perfectly intelligent people taping up mawkish posters of, say, a shirtless stud cradling a baby, or of dogs playing poker, Rebecca, true to herself, hung a black-and-white poster of Jordan, arms outstretched, a ball palmed in one hand—a cherished object in a room so small you could touch

51

all four walls while lying on her bed (which I was known to do). Occasionally she wore a pair of old socks emblazoned with Jordan's number, 23. The family had, among its VHS collection, two videos of Michael Jordan highlights. So when the United Center, Jordan's home arena, was dismantled, she and her twin sister had a small bit of the gym floor shipped to Austin, as a gift for Ben. On it was a scuff mark, one Rebecca likes to think was left by Jordan on his way to the hoop.

She wasn't sold that I could dunk. But in some ways I was strapping her to my back: Even as she endured my kvetching about having to work out, my stench upon my return, my groans about my cramping, massing muscles—this was the closest she would ever come to dunking, too.

328 days to go: *Morning sprints—6 x 25 meters; 4 x 50 meters; 2 x 100 meters, repeated twice—and evening lifting: Hamstrings nearly parallel when I squat now, even if I feel a bit like the Tin Woodsman. Squatting only the weight of the bar, 45 pounds. Three sets of 12 lunges, bearing 15 pound dumbbells. Two sets of 15 squat jumps. Weight: 192 pounds. Why does the carrot juice carton say "Not a reduced-calorie food"? Can I not even eat carrots? Jesus.*

Progress report: One night after work I decided to shoot around at a court down the street. A bit of relaxation after a muscle-bound week. Some guy dribbling around at the other end of the court challenged me to a game of one-on-one. He was my height, roughly, a bearded white guy in his mid-20s with some roundness at the middle. Cody was his name. He might have been me a couple of months earlier, a few pounds heavier and a few pounds less muscular. With much sweating, I had dropped more than 10 pounds and cut my body-fat percentage from 21 to roughly 17. Less to carry to the rim.

After he checked me the ball, I danced around the top of the key for a minute, trying to figure out whether just to shoot, lazily, or do my usual back-myself-in-and-awkwardly-post-up maneuver, like a Mack truck reversing slowly into a loading dock. Or, I suddenly thought, I could drive. This was an unusual thought for me, given my normal on-court deference and overall slow foot-speed. What the hell, I thought, and crossed the ball over to my left, lowered my right shoulder, and, in a couple of quick, bounding steps, beat him, surprisingly easily, to the backboard. Basket. I caught the ball as it came through the net and handed it to him so that, again, we could engage in the ball-check ritual. I wandered back to the top of the key, rubbing the sweat from my fogged-up glasses, and he bounced it back to me. I drove again: another bucket. Again and again, to the left or right. I was getting higher up, faster, even if I was nowhere close to the rim. I began wondering: Was he that weak, or was I suddenly that strong? Had these leg exercises, these glute crunches, these squats, improved my jumping already? I beat him 15–2. He wanted to run it back, so I ran all over him, again, winning 15–6. A masochist, he asked me if I wanted to play again. No thanks, I said. And stepped to the side of the court to do three sets of 12 squat jumps.

4

Taking the Measure of the Man

282 days left: *No pasta or beer or much else in the carbohydrate department now for two months. Can't say I mind it much: Down to 186 pounds. Today's regimen: Three sets of 20 squat jumps; 300 ankle hops; 150 split jumps, changing which foot lands on a box and which on the floor; 140 jump-for-joys, as I call them, the sort of jumps where you go as high as you might; three sets of 15 lunges, with a 15-pound dumbbell in each hand; three sets of two pull-ups—six more than I could muster in late August; three sets of 15 push-ups; three sets of 10 hamstring curls at 65 pounds a set; three 30-second planks. Just beginning to muscle up.*

A visitor in the summer of 1867 to the Maine town of Belfast, tucked into the upper end of Penobscot Bay, might have found a group of young men gathered around a single muscular gymnast near an old farmhouse barn. There, at the center, in the body of 17-year-old Dudley Allen Sargent, the visitor would have seen a well-sculpted youth who had perfected himself to fight in a war he was too young to wage.

A couple of years earlier, in the closing months of the Civil War, Sargent had read *Anatomy, Physiology and Hygiene*, an 1852 book by a now-long-forgotten Dr. Cutter that included this exchange: "Agesilaus, King of Sparta, when asked what things boys should learn, replied, 'Those which they will practice when they become men.'" The practice, of course, whether in Sparta or in 1864 Maine, was for battle. "To develop my body became an obsession with me," Sargent later wrote.

He took up dumbbells, Indian clubs, and boxing, and systemized his gymnastics, all the while undertaking heavy-lifting farm work, rising at 6 a.m. in the winter and 4 a.m. in the summer to complete his hay-baling, lumbering, and wood-chopping chores. He began hosting gymnastics and tumbling demonstrations at his family barn, and soon he was drawing followers. As a pastime, Sargent would lie on the barnyard floor and wrestle against two or three boys at the same time who were charged with keeping him down. The local newspaper, the *Republican Journal*, described him as a "young Hercules." Despite the press, the boys largely trained in secret, to insulate themselves from a frosty Maine sensibility that predictably turned against the gym work.

> I came across a public prejudice which I was later to know
> well and battle hard. Some parents forbade their boys to
> take part in any kind of gymnastics which they regarded
> as "monkey shines" and "gymkinks." The gymnasium was
> regarded in the same light as were billiard saloons and
> bowling alleys. Whether this ill feeling was a survival of
> a Puritanical spirit that tended to stamp out all manifes-
> tations of life and joyousness, or whether its cause lay in

the custom of the German turners of performing in cheap recreation halls and saloons, I cannot say.

Eventually, Sargent fled to the circus, literally, to become a trapeze artist and tumbler. The Goldie Brothers circus was wearying, of course—among other things, a tiresome senior clown always stole Sargent's buckwheat pancakes—and in less than a year he made his way back to Belfast, weighing whether he might go into law, the ministry, or medicine. A friend, returned from Bowdoin in the summer of 1869, told him about a possible vacancy in the directorship of the school's gymnasium.

Sargent was 20 years old, a former circus runaway with little money and no college degree. But he had shown himself to be a keen, serious student of the body in a period without any such discipline. Bowdoin took him on, and that fall, in his inaugural address, the president of the college, a Mr. Harris, told his charges:

> Other things being equal, the healthy man is the happiest and makes others the happier. He is the more pleasant husband and father, the more generous friend, the cheerer and helper of the sad, in every position and relation, the wholesome man. He radiates joy. Health, as the perpetual spring of animation and energy, is the first requisite of success. It must never be out of sight in the administration of a college.

The speech evidently made an impression on Sargent. As the gym janitor, instructor, and director, he gained the respect of the students—at one point challenging a series of them to boxing matches, which he invariably won—and remade the modest gym, importing the latest weightlifting mechanisms from Germany.

After a half-dozen years at Bowdoin, however, he went to Yale for a medical degree before moving on to New York, to open his own gym, at 24th and Broadway.

But even as he began training New Yorkers, he longed to return to the academy, to press upon young men his particular form of physical education, which conflated physical health and hygiene with moral rectitude. The teacher of physical education, a term promoted by Sargent, aims "to make the weak strong, the crooked straight, the timid courageous," he would write in one academic journal. "His aim is not only to keep them well and prevent disease, but to lift them to a higher plane of living, morally and intellectually, as well as physically." Such a teacher should combine technical ability and scientific know-how with "a great deal of the moral earnestness and devotion of the Christian minister."* He wrote to a host of universities advertising his services and urging them to consider the addition of physical education to their curricula. The letters back were discouraging. "For the most part, of course, these letters were polite and courteous," he later wrote, "but not one of them left a shadow of a doubt as to the absolute impossibility of establishing such a department in any of the colleges concerned."

Eventually, one school called upon him—Harvard. In 1879, he was named director of the university gym, a job he would hold for the rest of his life. The university became a platform for his search, through physical education, to discover the true potential of his

---

* The relationship between morality and physical robustness was a common thread in Sargent's time. "Even if the day ever dawns in which [muscular vigor] will not be needed for fighting the old heavy battles against nature," William James, the psychologist and brother of Henry James, wrote in his 1911 *Gospel of Relaxation*, "it will still always be needed to furnish the background of sanity, serenity, and cheerfulness to life, to give moral elasticity to our disposition, to round off the wiry edge of our fretfulness, and make us good-humored and easy of approach."

students. He dedicated himself to a field known as anthropometry, which sought to calculate, through a mass of tables, human capability.* In time, he gained a reputation, along with a beard and a serious mien, for his curriculum and his innovations, including a primitive rowing tank, "so geared that a four-mile boat race might be rowed in a hundred-foot gymnasium," as one of his protégés described it.

Prolific and dedicated, fixated on capturing the statistical measure of the man, Sargent late in his career grew "overtired and fagged out by trying to do too much," a friend wrote. But in 1921, at the age of 72, with death still several years off, Dudley Sargent, then a man with a long white beard that looked as if it belonged in the nineteenth century, made one final innovation: the jump test. That year, in the *American Physical Education Review*, he published a paper, titled "The Physical Test of a Man," that codified the vertical jump and gave it a sheen of scientific seriousness. "In popular estimation," he wrote, "it takes so many inches and so many pounds and a certain size chest girth to make a man." But there remained a mysterious "unknown equation" of athletic capability, one that ought to account for power and efficiency and that could be practically tested and numerically expressed. (After all, no one glancing at Spud Webb's chest girth would think he had the capabilities he did.) The jump test he conceived would reveal that previously unknowable relationship. It was a discovery "so self-evident that any fool ought to have thought" of it. "It is so simple and yet so effective for testing the strongest man or weakest woman or child that

---

* He derived, through some odd mathematics, a "vitality coefficient." In an 1887 treatise on the principal characteristics of the human body, Sargent describes the vitality coefficient as equal to the respiratory-height coefficient multiplied by the organic health coefficient. Got that? The same pamphlet, which gives directions for measuring abdomen depth, limb girth, and chest expansion—to determine said coefficients—offers health tips. "Rub vigorously after a cold bath; secure a fine glow, becoming thoroughly alive."

one feels almost like apologizing to the general public for mentioning it," he wrote. It was in that paper, in a sense, sprung from the cold-eyed, empirical confidence of the twentieth century, that the modern sports industry—the NFL combine, the poking and prodding of athletes, the statistical obsessions of baseball—was born. In it lies the notion that if you peer closely enough at a person, you can discern his true, mathematical composition.

The jump, wrote Sargent, required strength, speed, and something more ineffable, which he gropingly called energy.

> First, no one would deny that the ability to project one's weight 20 or 30 inches into the air, against the force of gravity, requires *strength* on the part of the muscles engaged in the effort. No one would deny that the effort would have to be made with a certain degree of velocity or *speed* in order to create impetus enough to carry the body twenty inches above its own level in the standing position. Further, no one would deny that back of the requisite strength of muscle fibers and rapidity with which they are made to contract there must be *energy*, "pep," "vim," vitality, or whatever it may be termed which drives our internal machinery. Overlapping all, of course, is the skill or dexterity with which the jump is executed. [Italics in original.]

After years of inspecting young men; of training them to make the most of themselves, very much as athletes as well as moral beings; of amassing reams of statistics on human potential, he settled on the measurement of the jump as the single telling factor of a person's physical makeup.

●

"Damn, I hate jumping," whined Jamal Carter, a 315-pound, dread-locked, baby-faced 23-year-old. No one was listening. Carter was seated on a corner of artificial turf, his squat legs sprawled out ahead of him, stretching with a bunch of other elephant-sized men, as he prepared for the vertical jump test at the Houston Texans' Method-ist Training Facility. These were, or hoped to be, NFL prospects.

Each year, the nation's top college football players are invited to Indianapolis to work out for scouts from each of the NFL's 32 teams. Then there are the regional combines like this one, for the second- or third-tier players. Ones like Carter, who had recorded a decent 21 tackles in nine games as a senior at Jackson State. Or guys who dropped out of college or were thrown off their teams or were five years out of school but thought themselves great sandlot play-ers. Teammates from Wichita, Kansas, had piled into a Hyundai Sonata to drive the 600 miles—one way. All of them had ponied up the $250 in admission and found their way to Houston with the hope of being discovered. All to be shouted at over a long morn-ing as they participated in the same tests and drills as the players who had a more realistic shot in Indianapolis. The tryouts included the vertical jump test, the one that Dudley Sargent had landed on to get the measure of the man. The stakes could not be greater for the players, who wanted to display their potential, and the teams hoping to divine it. On the line, in the players' minds, at least, were millions of dollars in contracts, and millions more, for the teams, wrapped up in jersey sales, television deals, and, ultimately, victories. Mike Hagen, a handsome, polo-shirt-and-khaki-shorts-wearing 50-something scout with the tired gait of an ex-athlete, called these tests "measurables." The underlying notion was that beneath all that raw flesh, scouts could latch on to the true potential of the athlete. "This isn't about playing football, this is about being an athlete," Hagen told me.

In what I took to be nervousness, the players were mostly quiet, almost solemn, as they stretched and prepared for a series of drills. The odds of success, measured in an NFL contract, for Carter and the 528 other athletes showing up in Houston—one of 10 regional combine sites—were slim. In 2012, 1,999 kids worked out at the regional combines; only four were ultimately drafted by NFL teams. "There is no unheard-of player under a rock," Sean McKee, a bright-eyed and leprechaun-like NFL official, told me. "The analysis nowadays is so deep and thorough. But everybody here thinks he's the one that's been under a rock—that his talent is undiscovered.

"We offer closure," he continued. "No one has ever looked these guys in the eye and said, 'It's not happening. Go get a job, support your family.'"

Sargent's vertical jump test—along with the 40-yard dash (also known as the "dash for cash"), the shuttle run, the broad jump, the bench-press test—would determine whether actual talent lurked among this crew of massive bodies. "It all correlates," Eric Lougas, a wide-shouldered, wide-faced football coach from Atlanta, brought in to help run the combine, told me. "Show me a guy with a good vertical, I'll show you that guy running a good forty-yard dash."

The only real scout on hand was the junior man from the hometown team. Chris Blanco, a Pumas-wearing Californian, a recently minted law school grad wearing a loose-fitting Houston Texans hoodie, stood with a clipboard, barely bothering to make notes as one would-be player after another tried the 40-yard dash. I asked him if I would be able to tell the difference if Andre Johnson, the All-Pro wide receiver for the Texans, were doing these drills. "No doubt," he said. "He just moves different from these guys. Totally smooth, totally efficient. Almost all these guys look like they're trying."

The players, wearing black uniforms—some sized as large as XXXL—rotated, over the course of a couple of hours, between the different workout stations. The place was oddly quiet, muffled like a mitten, despite the huge men stomping around. In the vast, hollowed-out belly of the practice facility bubble, they appeared to me as a thousand Jonahs, each mumbling a prayer for his salvation. So much—contracts, money, esteem—rode on the outcome of just a few tests.

Gathered in a corner of the practice bubble by the facility's concrete loading dock, one group, including Jamal Carter, prepared to jump. One at a time they stood next to a tall vertical staff hinged up high with a host of perpendicular rods that were ready to swing with the slightest nudge. This device is called a Vertec. The aim is to touch the highest possible rod. Eric Lougas, the friendly football official from Atlanta, stood above them on a stepladder, going through the instructions.

"First I'm going to measure your reach with your arm straight above your head," he said, striking the role of a benevolent teacher. "Then you're going to get two jumps. Don't swing the hell out of your hand, because if you do, what does that mean? It means you're using all your energy to bring your hand all the way back. Just tap it."

Soon it was Jamal's turn. He rolled his neck back and forth and peered up at the Vertec. Then he went into a half squat and rolled his way up: 18 inches. About my vertical, except he had at least 120 pounds on me.

Lougas used the stick like a croupier to reset the rods.

Next up was Tommy White, a handsome six-foot-tall, 210-pound cornerback who had graduated from Grambling two years earlier. He had told me he'd spent the last two years as a vegetarian while dedicating his life to working out and practicing yoga. He stepped beneath the Vertec and rocked down and up, down and up, down and up, and each time I was sure he would jump. Finally, he exploded into the air with a wail and snapped the rods.

"Thirty-eight inches!" Lougas boomed. "Now let's get forty! Get forty!"

The other athletes, none of whom had been paying attention, were now suddenly into it.

"C'mon, bro!" shouted one.

"Get on up!" shouted another.

Forty inches would turn heads among scouts. Calvin Johnson, the star wide receiver for the Detroit Lions known as Megatron, had notched a 46-inch vertical at the NFL combine.

Again, White went into his up-and-down, springlike gyrations, as if he were trying to fake us all out. Again, he yelped as he heaved upward. Once more: 38 inches.

I thought White would be pleased, but he looked deflated.

"I could do better," he told me.

That seemed to be a theme, to a man. Each of the athletes I interviewed claimed to have performed better in his own workouts, at his own gym. Disappointment lingered in the air, pressed out through the cold mechanism of assessment.

●

Presiding over this mass of men was Coach Stephen Austin, a small, trim tyrant. "We are guests at this facility," he told the players, assembled on bended knee before him. "If when I get to the restroom I see as much as a piece of tissue on the ground, there's going to be trouble. Demeanor, focus—we evaluate everything; it's as important as your athletic ability." Privately, I very much doubted that anything much mattered other than those "measurables" of athleticism. "This is serious business," Austin continued. "This is a day you're going to remember for a long, long time."

That part I believed. Besides participating in the jumping and sprinting tests and some football drills, before men in clipboards and polo shirts who unfeelingly jotted down notes that could shape their lives, the players had to strip down, one by one, and stand in front of a video camera, whose operator slowly panned up and down their bodies, front and back. There was a peculiar cast to the whole thing, of young men, almost all black, being examined by a group of largely white men, of being poked and prodded, and of being told to strip to their underwear so that a young white man could videotape their physiques. On the auction block, slaves had to, among other things, jump for their would-be buyers. A Sargent test before Sargent was even born, and another act of humiliation, surely, in a life of humiliation and worse. Beyond the degradation, it was meant, along with looking at teeth and musculature, as a crude way to test their potential. The film clips, I was told, would eventually be uploaded to the Internet, sortable by player, and I imagined team executives sitting before the glow of their computers, in a dark room, gazing at the physique of one man and then another.

At a lunch buffet reserved for the scouts and NFL staff, I started chatting with Austin. I asked him where he had been a coach. No, he told me, he was merely acting the role. "I'm a businessman," he said. "I play many parts." He had started his company in the late 1980s, one of many to scout out college kids as football became big business. "My big innovation," he said—and here I waited for something profound, some insight into the mix of qualitative and quantitative examination—"was pre-registration." I glanced from side to side as I tried to figure out if I was missing something. Evidently, little had been tabulated about the prior experience of the players—what college football conferences they had played in, for instance. So he kept and built up data so that players could be compared as apples to apples. I didn't quite understand, since a vertical is a vertical is a vertical, but I think his point was that the consequence of all this pre-registration was that he could distribute tables to general managers and football teams.

"You ever really read what I send you?" he told me he once asked Ozzie Newsome, a big man, a former football player, who had become one of the few black general managers in the league. They were in Newsome's office, and like a cloud passing on a sunny day, Newsome pushed himself, sitting in a swivel chair, to the side, and gestured toward a shelf of binders that were behind him.

"All the time."

Austin told this story with obvious relish: Newsome's Baltimore Ravens had become Super Bowl champions only a couple of weeks earlier, and Newsome had been credited with building the franchise through savvy scouting.

Evidently the key was pre-registration.

•

At the end of the day, the players assembled again on bended knees before Austin. They had carefully aligned themselves at midfield, between the hash marks, as they had been instructed by Lougas, one of Austin's underlings. "If you don't want to make Coach Austin mad, do as I tell you," he whispered to the athletes. These would-be players, it was clear to me, would do anything, small as it might be, to move to the super-regional. I knew Jamal Carter wasn't going to make it—I saw him icing his hamstring, which he had popped during the 40-yard dash. And Tommy White had rubbed Coach Austin the wrong way by taking a little longer than the others to get into a three-point stance before starting his dash: "What is this guy, a prima donna?" Austin shouted to no one in particular. Tommy compounded matters by complaining that the officials had not recorded his correct time—he insisted that he was consistently faster. (Of the dozen or so players I interviewed, only one, I would later learn, made it past the Houston event. He was invited to a Washington Redskins minicamp before getting cut.)

"You need to have a second dream," Austin told them. Then they heard a couple of spiels. An NFL officer told them about officiating opportunities. "Whether you make the NFL or not, there are other job opportunities so you can be involved in the game," she told them, touching on the necessity of clock operators to a group still unready to recognize the shortcomings of their own potential. "Officials maintain the integrity of the game. And you can be officiating the rest of your lives." The players remained politely quiet, even if many of them appeared to be staring into space. And then a chief with the Navy SEALs, a sponsor of the regional combines, spoke warmly about military camaraderie. He was tall, broad-shouldered, curly-haired, and wearing camouflage. "You guys have what the SEALs are looking for," he said. As the group finally disbanded—"We'll call you if we want you to come to Dallas," the site of the next round of tryouts, Austin told them—several of the players came up to the SEAL chief. "How much does it pay?" they wanted to know.

There was something admirable, touchingly sincere, and maybe a little desperate about these men. Most of us surrender our rock-star dreams at an early age. They still wanted to see what they could make of themselves even as the realities of the world bore down on them. Just as I was packing up to leave, a player came up to me, a slender, muscular kid. He wanted me to take a look at a DVD he had prepared of his football highlights; he hoped, I realized, that I was an agent or maybe a scout—another white dude with a notepad who might offer a bit of help. No second dream—not yet, at least.

Dec. 26, 242 days left: *Back to 188 on the scale. Oy gevalt. No big surprise: Christmas Eve dinner of cuminy pork loin in cognac cream sauce with pears; Brussels sprouts broiled with* *onion and bacon; Linzer torte, baked by yours truly, with fat dollops of whipped cream. Why should I run from my Austro-Hungarian destiny? Squatting 120 pounds, more than twice my starting weight.*

Progress report: Setbacks. A New Year's Eve scooter accident—braking suddenly on a rain-slick road—left me tumbling along the concrete like a stuntman; I was largely uninjured, save a bloody elbow and a bruised left hip. And then, three days ago, I tweaked my back while squatting. I wasn't lifting much—roughly the weight of a small woman lying across my shoulders—but I'm laying off the weights for a couple of weeks.

5

A Natural History of Leaping

In 2003, Malcolm Burrows, a professor at Cambridge who specializes in the nervous system of locusts, was leading some of his students in field research in the lovely English coastal town of Wells-next-the-Sea when he decided to ask them to capture some froghoppers, also known as the spittlebug for the frothy liquid in which their babies immerse themselves. They are wet, incontinent creatures. "There are constantly little drops of water coming off them," Burrows told me. "So when you're in the forests of northern Canada and it feels wet, that's because they're peeing on you all the time." Burrows had read recently about the jumping exploits of fleas, and he wondered how the common froghopper, which he had noticed quickly flitting about, stacked up. Students being students, they obliged. Armed with narrow-holed butterfly nets, they managed to capture some of the bugs. At the lab, Burrows put one of the froghoppers in front of a high-speed camera that captured 500 frames per second. The first frame spotted the insect; the rest were blank. After he upgraded to a 5,000-frames-per-second camera, only five frames caught the creature.

It turned out that with a combination of an ingenious catapulting mechanism and an advanced energy-storage-and-release system, the froghopper had jumped incredibly far incredibly quickly. "Froghoppers have relatively short and light hind legs that are powered by huge jumping muscles and a novel locking mechanism that allows force generated before the jump to be released rapidly," Burrows later wrote in the journal *Nature*. When it prepares to jump, the froghopper pulls its legs in and they lock in place by means of Velcro-like pads between them. As the massive muscles slowly contract, the insect stores the potential energy in its very skeleton, in a pair of curved structures shaped liked archery bows. That skeleton is composed, in part, of a rubberlike material called *resilin,* known for being the world's most efficient elastic material. Before each jump, the skeleton bends and bends and then—*thwap!*—it releases, the legs extend, and the froghopper goes flying.

The froghopper, Burrows explained, combined and exploited the best biomechanics of other creatures, making it a world-champion jumper. "There are two basic body designs for jumping that enable many animals to escape from predators, to increase their speed of locomotion or to launch into flight," Burrows wrote in his *Nature* article. "Animals with long legs (bush babies, kangaroos and frogs, for example) have a levering power that enables them to use less force to jump the same distance as short-legged animals of comparable mass, whereas those with short legs must rely on the release of stored energy in a rapid catapult action." (Put us warm-blooded humans in the bush babies' category: That's why champion high jumpers tend to have long legs, to use as giant levers as they flip themselves skyward.)

The froghopper has "developed the most amazing mechanisms for jumping," Burrows tells me. The froghoppers exert a force roughly 400 times their body weight on the ground, about three times as much as your average flea, and 150 to 200 times as much as a human. Another way of thinking about it: If an adult human could exert that kind of force when jumping, she could clear the Gateway Arch in St. Louis.

•

To visit Burrows, I boarded a 9:45 train one bright weekday morning at London's King's Cross Station. It was a clear day, recalling the London "spread out in the sun / Its postal districts packed like squares of wheat" that Philip Larkin described in his train-ride poem "The Whitsun Weddings." Rebecca and I were visiting England chiefly to see her brother Ben and his family. Besides novel-writing and teaching, Ben has carved out an essayist's life as the go-to guy among English periodicals for something literary-minded about the sporting scene. Forty years old, and the man can still throw it down. But what with the responsibilities about the house and his teaching obligations and the shortage of basketball rims around London, he seldom finds himself on a court nowadays. Even so, when he and I play—I in the best shape of my life, he in average shape for him—he can still cruise by me as if I've stopped to lace my sneakers.

Eleven minutes after pulling out of King's Cross, I saw my first horse in pasture. Happy horse, happy pasture, very England. "All windows down, all cushions hot, all sense / of being in a hurry, gone," wrote Larkin, looking out on a similar landscape, and sitting there I felt a momentary reprieve—a sun-basking exhalation—from the constant haste of my upward rush. As we sped by the backs of houses, the tableaux had the distant ring of familiarity for me. As a 13-year-old trailing my academic parents, and then, later, as a graduate student myself, I had done my time in England. An hour later, after the train pulled into Cambridge station, I stepped out to the meaty, buttery smell of a platform vendor's Cornish pasties, and long-suppressed memories of execrable English public-school food wafted through the olfactories of my mind. I walked up the High Street, past a kebab joint with menus in English, French, Spanish, and Korean plastered to the windows, and across the way, a World War I memorial—a dutiful young soldier, forever striding with rucksack and rifle, a grim, defiant expression wiped across his thin-lipped face.

Burrows's office was in the zoology department, just off the sort of pretty, winding cobblestone street that English university towns do so well. The shelves were jammed with vials, each containing a dead insect, as if suspended in mid–buzzy leap. I had come hoping to learn something about the elusive mechanics of jumping, and about why some species had been selected for their leaping ability. For years, Burrows had performed locust surgery, perfecting, like one of those artists who paint landscapes on a grain of rice, superfine movements with forceps to implant electrodes on nerve cells only 30 microns in diameter, about the size of a very fine human hair. But promoted to department chair, where he had to busy himself with the muck of academic administration—faculty complaints about parking permits and the like—he fell out of practice. By 2003, his best days of locust surgery behind him, he decided he needed something new and engaging. That's what brought him to the froghopper.

The insect was selected for its jumping prowess either because jumping is a more efficient way to get from one tree to the next than crawling down and back up again, or because it's a very fast way to "get away from nasty things that want to eat them," Burrows told me. Jumping gets you out of a bad situation faster than flying, which requires the beating of wings to take off—imagine how long it takes a helicopter to get its blades up to speed before takeoff.

How fast is the froghopper's jumping action? An average human being's top reaction time is 200 milliseconds—that's how long we need, minimum, to recognize and dodge a baseball headed for our face or, put more poetically, to identify emotions in the countenance of a fellow human. The froghopper can jump in about one millisecond. So fast, Burrows told me, that something beyond muscle contraction must explain its jumping ability.

I was interviewing Burrows at his desk, which is tucked beneath a window in one corner of the lab. In front of me, in some vials, sat some insects. They were froghoppers, he said. Still alive. He had captured them the previous evening with his eight-year-old grandson in his backyard garden. He took down a butterfly net—a short-handled one with an old, beaten canvas bag attached, its holes minuscule—and, in a brief hand ballet, gracefully whooshed it this way and that along the ground in a demonstration.

"May I tap them?" I asked, nodding at the vials, as if I were a kid wondering about banging on an aquarium glass.

"Of course," he said.

I tapped—and not much happened. Just like at the aquarium.

Minus the specs, maybe it was the Clark Kent of insects: mild-mannered, vast powers concealed.

The discovery of the froghopper's jumping talent and technique put Burrows's career on a new trajectory, so to speak. His mind captured by the ingenious jumping mechanisms, he started to look at how other insects measured up as he sought to secure the frog-hopper's title of highest jumper. He filled books and binders with notes—the spine of one reads, simply, "Jumping Cockroaches." In his lab, he set up a mini-theater to measure and record the insects' jumping. They were placed in a small glass box, brightly lit by a half-dozen or so lamps, as if on a stage. A foot away sat a video camera capable of taking 100,000 frames per second.

He showed me some of the films. They were beautiful. A lace-wing tumbling through the air, in floaty slow motion, and an exqui-sitely defined silhouette of a praying mantis, perched at one end of its stage as it prepared to leap—and then did, its long abdomen curl-ing forward in a show of fingernail-sized core strength. It was the twenty-first-century equivalent of the Muybridge photos. I could have watched for hours. It was this film work that clued Burrows in to the coordinated mechanisms that explained the insect's spec-tacularly quick jumping ability.

When I saw him, Burrows was excited, in his modest way, about a forthcoming article in *Science* (something most scientists hope for their entire careers), about a juvenile insect that jumped with the use of gears, the first detected in nature. The insect had developed a mechanism of interlocking gears, like those in an old wristwatch, to make sure that its legs sprang simultaneously.

I asked him if an individual insect could perfect technique, if it had the smarts and creativity to improve what it was born with. If, through practice or insight, it could stretch the evolutionarily determined radius of its species' abilities.

"I shouldn't think so," he said.

As I was leaving, Burrows asked what I knew about the Fosbury Flop. Only that it's the standard high-jump technique, I said. He smiled a small, knowing smile. As a young post-doc out of Scotland, Burrows had been doing research at Oregon when he decided to attend one of the school's famed track-and-field meets. It was there, at a track meet in 1967, that Burrows saw a pale whippet of a man, from the state school down the road, using an unorthodox, nearly backward high-jumping technique, flinging his head and shoulders back-first, his belly to the sky, to clear an ever-higher bar. This man, in short, was doing what none of his prized insects seemed capable of—jumping beyond his apparent mechanical limits with sheer ingenuity and unorthodox technique. The man was Dick Fosbury.

●

From the armswing to the stretch of the Achilles, jumping is surprisingly complicated for such a simple maneuver. Biomechanically speaking, the goal of the jump is to raise the center of mass following foot-to-ground contact. So the upward force at the feet has to be greater than body weight to produce upward acceleration and liftoff. Of course, Sargent was right: Whatever force you generate in the first stage of the jump is followed by the decelerating effect of gravity while you're in the air. Remember, you can't add force once you're airborne, because your feet are no longer touching the ground. "As our upward velocity and consequent shortening velocity of the muscles increase, the ability of the muscles to produce force decreases," Arthur Chapman writes in his textbook *Biomechanical Analysis of Fundamental Human Movements*. "In this sense jumping is a self-defeating activity." (Good old Arthur, always there with an encouraging word.)

Force, you might recall from basic physics, is mass times acceleration. A big dude like me needs more force to generate an equal amount of acceleration as a little guy. Acceleration is key to jumping—it's why Todd Wright is fond of saying that he wants his players to push through the ground *as quickly as possible*. And that quickness comes from those fast-twitch muscles. That's where the plyometrics and the weightlifting—the sweat equity of this endeavor—are meant to help. Stronger muscles, recruited more quickly, mean more force applied to the ground, faster. Stretching is crucial here, too: Just as you can get more force from a hammer if you swing it over a greater arc, you can push through the ground with greater force the higher up you can bring your knees when bounding toward a leaping jump. The bigger the motion, the more opportunity to apply force: Work is force times distance, and power is work divided by time. "The less ankle flexibility you have, for example, the shorter the arc through which you will be able to apply driving force against the ground," John Jerome writes in *Staying Supple*, his treatise on flexibility. "With a briefer arc, you'll expend more energy to attain the same speed."

Directing myself upward as efficiently as possible will also require some old-fashioned improvement of my biomechanics. Newton's third law of motion holds, famously, that "action and reaction are equal and opposite." The recoil of a fired handgun, for example. To get a sense of this, get on your bathroom scale and quickly raise your arms above your head. The scale, you'll notice, will spin rapidly as if you've grown in weight. In effect, you're jumping, even if your feet haven't left ground. That's because the muscular force required to push up your arms is also pushing the rest of your body down toward the earth. The reverse, of course, is also true: Get on the scale with your hands above your head. Now rapidly drop your hands by your sides. The scale, as your arms move through the air, will lighten. You're falling, now, as your body pushes force skyward. Thus, the first lesson of biomechanics: The armswing is an important part of the jump. (These things have been examined closely: Jody Jensen, a University of Texas exercise science professor, titled her PhD thesis "Intersegmental dynamics: Contribution of the arm-swing to propulsion mechanics in the vertical jump." Seriously.)

With roughly eight months to go in my dunking year, fundamental questions involving the physics of explosion remained ahead of me: Should I, for instance, undertake the running one-foot leap or the quick two-foot plant-and-jump? Some dunkers, like Jordan, are known for their running leaps; others, like Amar'e Stoudemire, a forward for the Knicks, typically jump off two feet. That difference gets at the idiosyncrasy of human movement even as we're trying to reach peak performance: Why, for example, is there not a single, obvious way for a baseball player to hold a bat as he awaits a pitch? Professional golfers talk about tinkering with their swing, as if there isn't an empirically, precisely correct style. When I played Little League baseball on Manhattan's Upper West Side, sandlot fields where dads tossed out kitty litter to soak up unwanted puddles and where both teams carefully walked before each game to pick up shards of broken glass, I modeled my stance after that of Darryl Strawberry, my hero on the New York Mets, who pulled his tall, skinny body in tight, like he was stuck in a crowded elevator. He held his bat up and back and erect, close to his shoulder, as if he were in a color guard, bearing the American flag, churning the bat in small, menacing little vertical circles as he awaited delivery of the pitch: the straw that stirred the drink. That stance was just about the opposite of the pose struck by his teammate Lee Mazzilli, he of the matinee-idol looks, who trailed his bat, limply, almost indifferently, behind him, hanging it just parallel to the ground. He won quicker bat speed, perhaps, because he didn't have to bother dropping his wrists as he swung. But he lacked Strawberry's power. A friend of mine, a baseball scout, tells me that the variety of stances and pitching motions comes down to the fact that "bodies are different." "Tall pitchers throw differently than short pitchers," he said. "Guys with quick arms—looser tendons, better fast-twitch muscles, et cetera—have different mechanics than guys with slower arms." That must be part of it: "Nature and human life are as various as our several constitutions," Thoreau wrote. But I suspect that body movement is like language and accent: You wring out, like a sponge, the steady stream of material and activity in which you soak. The idiosyncrasies of body movement are true in everyday life, too— far removed from the peak moments of hitting a ball or going up for a dunk. Even in cases with an agreed-upon, ideal, set way of doing things, we humans naturally deviate: In elementary school we painstakingly learn to write in print and script, along a sheet of horizontal dotted lines; still, as if we had fallen from a Tower of Babel of penmanship, we have among us a thousand ways of writing (to the extent we handwrite at all anymore). If Leonardo da Vinci's Vitruvian Man exemplified what our ideal proportions and postures ought to be, we rarely resemble his noble form. You can tell that just by looking at the postures or sitting positions of any dozen people on the subway. (Do you cross your legs or not? If so, right over left or left over right? At the knee or at the ankle?) The sloppiness of human idiosyncrasy doesn't help when you're trying to dunk a basketball. And all my kinks that Polly and Jamie had identified many months before had to be slowly unraveled. Physical habits had to be unlearned as I rebuilt myself. Not such an easy task.

•

Dick Fosbury always loved track and field, but the thing was that he was not all that good at it. Tall, thin, awkward—"a grew-too-fast kid," his Medford High coach would say—he chose the high jump, an event that, like the dunking of a basketball, favors an explosive jump. He had cleared 5'4" in junior high, a decent height, by means of the obsolete scissors technique—the athlete, like a sped-up can-can dancer, throws one leg and then the next over the bar. But then his progress stalled: to all appearances, he was another earnest but unremarkable young athlete. In high school, his coach pressed him to use the conventional jumping technique of the period. Known as the Western roll, or the belly roll, it requires the jumper to leap over the bar as if mounting a horse—facedown, flinging one leg over the bar and then the other. But young Fosbury did not take to the Western roll. Even at 6'4", he found himself struggling to match heights that were routine for his shorter teammates. He was not advancing—using the Western roll, he was backsliding, actually, failing at one meet to clear the opening height of 5'3". As the team made the trip to yet another track competition, in Grants Pass, a tiny town in southwest Oregon, a frustrated Fosbury, just a 15-year-old ninth-grader, told himself that if he couldn't surpass 5'4", he was through with the sport.

And so it was at a track meet in 1963 that a gangly boy began quietly tinkering with his cherished scissors technique. After matching his old junior high mark, he began to improvise: Instead of keeping his upper body erect, or even leaning slightly forward, as his legs scissored over the bar, he began throwing his hips higher—and his shoulders started falling backward in reaction. On his next jump, Fosbury cleared 5'6", pleased, if slightly befuddled, by the ease with which he suddenly had made a new height. Still, he was mostly ignored as other jumpers continued their warm-ups. Then, flinging his shoulders back a little farther, as if he were laid out in a chaise longue, he cleared 5'8". Now Fosbury had won onlookers, especially coaches. He was doing just about the opposite of the belly roll. He was belly-up! And instead of a foot being the first thing to clear the bar, it was his head and shoulders that were leading the way. Finally, the bar was set at 5'10", and Fosbury, his shoulders fully thrown back, as if he were lying flat on a flying carpet, floated over. At a single meet, he had achieved a full half-foot improvement over his previous best mark. "Fosbury flops over bar," read a newspaper photo caption back in Medford. And suddenly the Fosbury Flop was born. Just a weird oddity? Perhaps. Most high jumpers stubbornly stuck with convention, but Fosbury, who would study engineering in college, continued to tinker with his method. The brilliance of the flop was that it allowed him to clear the bar without necessarily lifting his center of mass (the point on the body where the mass above and below average out) over it; with his pelvis uplifted but his arms and legs dangling on either side of the bar, as if bridging in midair, his center of mass could technically sit below the bar even as each bit of his body, part by part, topped it.

Fosbury, it would turn out, was not merely a daring teenager with a quirky innovation: As he grew into his body, he grew into his technique. In 1968, a scant five years after nearly quitting the high jump altogether before backing into the flop, he earned a spot on the U.S. team for the Mexico City Olympics. Pitted against competitors who persisted in the Western roll's belly-down method, Fosbury, with his odd technique—not to mention a suspense-building habit before each attempt of clenching and unclenching his fists, sometimes for several minutes—endeared himself to the foreigners. In the end, to cries of acclaim, he delivered, taking gold with an Olympic-record height of 7'4¼".

There are two ways of looking at Fosbury's innovation. The prosaic way is that his breakthrough was possible only because of the evolution of the landing material on the other side of the bar. For decades, piles of sawdust and wood chips offered meager cushioning to those launching themselves skyward. But just as Fosbury got going, equipment manufacturers introduced primitive foam-landing pads, making it at least conceivable to a scrawny Oregon teen that he could throw himself over the bar and land on the back of his neck and shoulders.* Regions of the world that lack the foam landing pad continue to use bygone methods of high jumping. (High school jumpers in rural Kenya, lacking any cushioning surface save the ground itself, favor the Eastern cut-off, a contortion of limbs and twists that last garnered a world record in 1895. In fascinating YouTube videos, the jumpers land on their feet, keeping them happily safe.)

Another way to think about Fosbury's breakthrough, the way I prefer to think about it, is as a story of individual creativity. Even if Fosbury didn't realize, as a 15-year-old in a proto-hippie enclave in southwest Oregon, exactly what he was doing—that, in a bit of

---

* There was widespread trepidation among high jumpers those first couple of years at imitating Fosbury. In the *High Jump Book*, Dwight Stones, Fosbury's successor as the U.S. high jumping king, recalls athletes and coaches of the late 1960s worrying that the flop "would foster a generation of broken necks and damaged vertebrae . . . I was reluctant to intentionally adopt an unorthodox method of jumping if the injury potential was greater than with traditional jumping." But the chance of victory is persuasive: By the Munich Olympics, four years after Fosbury won on the world stage, 28 of 40 high jumpers used the flop. Today, virtually all elite jumpers use the method.

genius, he had figured out a way to yank himself above the bar even while leaving his center of mass below it—even if he didn't realize all these things off the bat, he had the chutzpah to describe a new kind of parabola. Fosbury had more or less instantly rewritten his potential, from a kid with pedestrian capabilities to one for whom a world record was within reach. Some sort of jumping ability was always there, deep inside Fosbury, but it needed some creative courage to reveal itself.

There are some lessons here for the would-be dunker: the fast run-up; the placing of the plant foot slightly ahead of the body to reduce forward momentum; the emphasis on pushing against the ground to drive yourself up (the harder you push against the ground, the harder it pushes back up against you, as Sir Isaac Newton, renowned for his jumping facility, put it); the possibility, finally, that a tall, skinny kid of no great apparent talent might find himself achieving greater heights than he thought possible.

But try as I might, I can't drum up a novel way to dunk. While Fosbury ingeniously figured out a way to hoist his entire body over an ever-higher bar that day in Grants Pass, he wasn't necessarily leaping any higher. I, on the other hand, do have to jump higher than I ever have before. As a human projectile keen on dunking, once I leave the ground there is little I can do to change my trajectory.

•

Jumping is safe, warned Arthur Chapman in his biomechanics textbook, unless it is done "in a room with a low ceiling." Yes, Arthur, again, thanks very much for that. For some guidance on form, and to practice in a space with higher ceilings, I headed to San Marcos, just a half hour south of Austin, to the warehouse-like gym of

Charles Austin. Austin is no ordinary jumper: As a 29-year-old, he won the gold medal in the high jump at the 1996 Olympics in Atlanta. He cleared 7'10". Put another way, he lifted his entire body over your front door. That mark remains the Olympic record. His gym, So High Fitness, is plastered with life-size photos of Austin preparing to jump, soaring, and bowing his head to accept another medal. Can't blame the guy. He's still competitive—cocky, even—and muscly; he still looks like he could clear at least 7 feet. In fact, he tells me he could. I ask him how often he works out now: "Just enough to maintain this," he says, glancing down at himself admiringly. He tells me with obvious pride and sincere anxiety about how his sons will overtake him. His middle son, Allex, started out gangly, clumsy. Now he's on the track team at Baylor. "He's a bigger, better version of me," Austin told me. "I hate telling him that, and I don't so often." Last year, before heading off to Baylor, Allex won the state high school high jump contest with a mark of 7'2".

" 'Dad, I don't understand why your records are so impressive,' " Austin says Allex told him.

" 'Clown, what you talking about?'

"He tells me: 'I won state twice in a row.' I said, 'You kidding? You think I care about state?' "

He laughs—he laughs at just about everything—as he tells this story. The youngest of 10 children, Charles Austin had grown up poor, black, and fatherless in the small, steamy South Texas town of Van Vleck. ("One thing that bothered me was seeing dads there cheering on their sons," he says, clearly angered still at his father's absence. "I know I'll be there for my kids.") His mother raised Charles and his siblings on a maid's salary. Part of his own narrative as a disciplined self-starter is how he became a student of track and field—without hiring a coach. "Most people paid crazy money," Austin said. (As he told me this, the irony of my paying him $100 an hour for a few training sessions wasn't lost on me.) "I sat there and watched other people working out. I picked the best sprinter, I looked at the arm action, how the feet strike the ground. How relaxed the hands are and whether the wrists are tight."

At the end of my first interview with Austin, he asked me to show him what I could do. "You can touch the rim, right?" I was nervous. I'd have to jump for one of the world's top jumpers. I hemmed and hawed. "Well, yes, but, you know, I'm not warmed up. I haven't stretched or anything." He smiled broadly. "Let's see it." So we stepped out of his office and onto the adjacent basketball court. I put down my pen and pad, squinted at the hoop, leaned backward to size it up, and then took a short run-up, leapt, and—managed to touch the rim. I was relieved.

Then I turned to see Austin doubled over, chortling. "If you can touch the rim cold, with that kind of form, I can get you to dunk," he said. It was a backhanded kind of encouragement, of course, yet I couldn't help laughing with him. I gathered my notebook and prepared to go. He stepped closer to me and leaned in conspiratorially. He told me that one reason he likes to work out is that he likes to "look nice," especially for his wife. He said he likes it when she comes up behind him and feels his strong, naked buttocks. He had a big grin on his face now and, not knowing what to say but feeling a need to say something, I nodded: "Yeah, but I've got some hairy buttocks." He stopped smiling, lightly pursed his lips, and nodded sagely back.

6

The Rise of the Dunk

Feb. 4, 202 days to go: *Weight is down to 185, emerging from post-Christmas funk. Body fat at 15 percent, down from 21 percent last August. Breakfast is light now: Grapefruit and a bowl of nonfat Greek yogurt, with just a teaspoon of black cherry jam mixed in. And roughly 21 pieces of fruit: Apples, prunes, grapes, bananas, and mangos. Squatting 135 pounds, carefully. The usual jumping exercises, and lunges with 17.5-pound dumbbells. Three sets of five pull-ups. Night-time sprints, beneath the buzzing orange magnesium lights of the deserted middle school track a couple of blocks from my house; 10 200-meter dashes, two minutes' rest after each rep. (The middle school, named for a Confederate soldier, counted Ann Richards as a teacher before she began her political career. I summon her steel.) Then four sets of 30 squat jumps. The heads of imagined book reviewers float before me, urging me on: I'm becoming convinced, in the cold heat and loneliness of late winter wind sprints, of a rela-*

tionship between my leaping ability and the quality of my writing.

> One afternoon during the summer, a bunch of us were playing against some older kids in the Carver schoolyard. I had grown to six-foot-eight but the older guys were still beating me up every time I got close to the basket. And whenever I called a foul, they'd say, "Shut up, bitch. I barely touched you." I got so mad that the next time I got the ball I just jumped as high as I could and stuffed the ball through the hoop. Everybody just stopped and stared at me. Then somebody said, "Do you know what you just did?"
>
> "What? What? What?"
>
> "You just dunked the ball, man!"

Thus begins the career of Chocolate Thunder, also known as Darryl Dawkins, a mountain of a man who gained fame in the late 1970s for shattering backboards with his tremendous dunks. As part of my dunking research, I had become a connoisseur of the basketball player autobiography, a genre in the universe of sports books distinct for its meditations on sex, race, and hardship. I would skim them quickly, flitting past stories about broken homes, absent fathers, and how Mormon girls were surprisingly easy. I learned to roll my eyes at another claim by Wilt Chamberlain that he concentrated at basketball just as he concentrated at sex because he wanted to be the best at whatever he did. I was looking for a single trope: the moment of the first dunk.

I wanted to catch that early spark that lit the bright flame of their careers. From our perspective, basketball players appear to be naturals, given their size and the ease with which they explode to the basket. But what is it like for them? Accounts of their first dunks provide a way for these players to wrestle with the larger question of how much of their ability was natural on the one hand—something of which they are reasonably proud—and how much was the result of sweat and effort on the other. The latter explanation fits into a narrative, often moving, about growing up in, and striving out of, poverty—especially black poverty—in America.

In Charles Barkley's autobiography, his mother, a husbandless maid who cleaned the homes of well-to-do whites and worked in the school cafeteria, recounts his persistence with a jump rope she bought him: "I thought he was going to jump that rope to death." Still, he needed to work hard to build up his legs to make the high school basketball team, "so one day I decided to try and jump over the chain-link fence around our house from a standing start, back and forth, like a jumpin' jack . . . Just being able to clear a three-and-a-half-foot fence was like a high, a rush," Barkley writes.

> "Man, this'll work," I thought.
>
> So, I did it again. And again.
>
> I didn't care how dangerous it was, but my fence-jumping used to drive Granny crazy. She tried to warn me that if I made a mistake and missed, I could mess myself up for life, "as for having children you know," she used to say. So she'd sit on the porch and watch me, like sitting there was going to make a difference if I fell. She sat there watching me for two or three hours, back and forth.

His first dunk, which came his senior season, was not only the moment when he lost his "basketball virginity" and "reached manhood"; it was also something more profound, a signal that he could leap out of the context in which he was born, a ticket out of impoverished Leeds, Alabama. "That's also when I started looking at basketball as a way to get to college. I knew my family would never be able to afford it, but it wasn't until I made that dunk that I thought I would have an opportunity to win a college scholarship."

You can see, in these stories, how the advent of dunking marked a shift in the game itself—a generational change, a racial change, a change that valued creativity as much as mechanics. In *Wilt*, which carries the subtitle *Just Like Any Other 7-Foot Black Millionaire Who Lives Next Door*, Chamberlain remembers that his high school coach, in the 1950s, didn't know what to make of him after his first dunk. "You should've seen the look on his face. The old coach looked like I'd just jumped through the basket myself, feet first." John Wooden, the legendary straitlaced UCLA coach (in fact, he was so obsessive that he taught freshmen players how to tie their shoes properly to avoid blisters), discouraged his players from dunking. "That's not necessary," he once told an interviewer. "If I want to see showmanship, I'll go see the [Harlem] Globetrotters and I'll enjoy it."

The dig was an enduring one, that black players somehow valued style over substance. In *White Men Can't Jump*, a worked-up Woody Harrelson tells Wesley Snipes's Sidney, "You're just like every other brother I ever met. You'd rather look good and lose than look bad and win." Dunking, to put it bluntly, was seen as a black way of scoring. The charge leveled stodgily against the move was one of showboating, as if the dunk couldn't be both beautiful and practical, entertaining and competitive. After the NBA finally integrated in the early 1950s, white sportswriters began railing against the dunk because it turned basketball into "aerial dogfights." No wonder: "The remaking of basketball in the shape of the African American aesthetic," the sociologist Gena Caponi-Tabery has written, "is an obvious case of subversion of the dominant culture by subordinate African Americans."

At the time Chamberlain stupefied his high school coach with a dunk, the sport was only a half-century or so removed from

when the "basket" of basketball was an honest-to-God peach basket In a gym at the International Training School for Students in
snowy Springfield, Massachusetts. In December 1891, charged by
the YMCA superintendent with inventing a game for young men
("a class of incorrigibles," the superintendent called them) that they
could play indoors by newly invented electrical light, James Naismith, a 30-year-old Canadian who had recently moved down from
Montreal, readied the gym for his new game of "basket ball." He instructed the school janitor to nail two peach baskets to the lower rail
of the gym balcony, which doubled as the running track. That rail
happened to be 10 feet off the ground—a completely minor, semi-
arbitrary architectural fact that, 125 years later, endured as the bane
of my existence.

The first game, nine versus nine, finished with a 1–0 score—no
word on whether that was a dunk. Basket ball was a hit from the
get-go, and it quickly spread through the YMCA network. By 1898,
the first professional league had formed, and in 1909, the first international game was held. Naismith went on to coach at Kansas, and
one of his players, Phog Allen, would become Kansas's most famous
coach. By Allen's coaching heyday, in the 1940s, white dunkers
started making names for themselves. The first of these was George
Mikan, who began his career at DePaul as a clumsy, bespectacled
oaf before learning to dominate inside. His greatest rival: Bob "Foot-
hills" Kurland, a seven-footer from Oklahoma A&M, who dunked
early and often. His was such a presence around the basket that officials devised the goaltending rule, specifically designed to prevent
Kurland from grabbing opponents' shots off the rim. Allen, maybe
because he didn't like that Kurland played for a rival, maybe because
he was old-fashioned, groused about dunking. "He would say the
game was overrun with 'goons,' as he called them," Matt Zeysing,
the basketball Hall of Fame archivist, told me. The crankiness was

common. "Basketball is for the birds—the gooney birds," the venerated sports columnist Shirley Povich wrote in *Sports Illustrated* in 1958. "The game lost this particular patron years back when it went vertical and put the accent on carnival freaks who achieved upper space by growing into it. They don't shoot baskets any more, they stuff them, like taxidermists."

By that point, Chamberlain was an undergraduate at Kansas, and dunking was fast becoming associated with blacks. Allen tried to get the rim pulled up to 12 feet in the 1950s, even using the higher rim in some exhibition games. Chamberlain himself, whom Allen recruited to Kansas in 1956 just before retiring, is said to have dunked on the 12-foot pilot rim. (In 2007, Dwight Howard, then the center for the Orlando Magic, dunked on a 12'6" basket in the NBA Slam Dunk competition.) As with Kurland, basketball officials changed the rules to put a stop to one of Chamberlain's patented moves: Just as he released a free-throw attempt, he would run toward the basket and jam home any miss. Now a player cannot cross the free-throw line until his or her shot has struck the rim or backboard.

Thwarting dunking became a theme of rule-making during the 1960s. Looking back, a half-century later, to a period in which white Americans were grappling uncomfortably, to put it mildly, with an expansion of civil rights, racial anxiety seems the obvious subtext to all the grumbling about dunking. It was a maneuver increasingly connected to black exuberance of the type white officials wanted to suppress. The slam dunk gave the players an opportunity to momentarily seize power and express outrage, Caponi-Tabery observed. When Lew Alcindor, later known as Kareem Abdul-Jabbar, showed up at UCLA to play in Wooden's program, such was his height and wingspan that he upended the established order with his dunking ability. His first year at UCLA, he led the freshmen to a 16-point victory over the varsity team, then the national defending champions. In 1967, at the start of Alcindor's junior year, the NCAA banned dunking. The stated reason was to save athletes from injury and to protect rims from damage. "Clearly, if I'd been white they never would have done it," Jabbar later wrote. "The dunk is one of basketball's great crowd pleasers. And there was no good reason to give it up except that this and other niggers were running away with the sport." Muffling Alcindor was just a pretext for a wider under-cutting of black athletes, Hunter College assistant basketball coach Robert Bownes told Pete Axthelm for his 1970 book *The City Game*. The dunking ban "wasn't put in to stop seven-footers," Bownes said. "It was put in to stop the six-foot-two brothers who could dazzle the crowd and embarrass much bigger white kids by dunking. The white establishment has an uncomfortable feeling that blacks are dominating too many areas of sports. So they're setting up all kinds of restrictions and barriers. Everyone knows that dunking is a trade-mark of great playground black athletes. And so they took it away."

The chief instigator of the ban was Adolph Rupp, whose all-white University of Kentucky team had faced the all-black squad from Texas Western in the 1966 championship. Rupp, who called blacks "coons," forbade his own players from dunking. Not Don Haskins, the coach of Western. Early in the game, one Kentucky would even-tually lose, Western's David Lattin dunked the ball on future NBA coach Pat Riley. For Rupp, the moment was symbolic as much as it was real. In the following off-season, "Rupp was so disgusted that he went to the NCAA rules committee and had the dunk banned from college basketball for 10 years," Lattin told Caponi-Tabery. "That one dunk, can you believe that? He was such a powerful coach, a big figure, that he had the dunk taken out of college basketball for a decade."

Until the NCAA's decision to ban it, the dunk was seen as a sure thing, a business-as-usual put-away move. With its ban, and as part of a newfound flair in 1970s professional basketball, the dunk be-came a piece of underground entertainment. Among its practitioners was a Long Island kid known for his enormous Afro. Starting at age eight, Julius Erving would take four steps at a time up and down his building. "Back then, before I was physically able, I felt these different things within me, certain moves, ways to dunk," he once told *Esquire* magazine. "I realized all I had to do was be patient and they would come. So I wasn't particularly surprised when they did, they were part of me for so long." By the time he was 14 he could dunk on full-height baskets. He was a contemporary of Alcindor's, and so, in the three years Erving played at UMass–Amherst, he was not allowed to dunk in games. (Alcindor, a very tall man, cast an even longer shadow.)

With the promise of greater pay, Erving, now known as Dr. J, passed up the NBA to play for the Virginia Squires in the upstart, freewheeling American Basketball Association. The ABA, led by commissioner George Mikan, the original dunker, put a premium on athleticism and showmanship; "above the rim" play instead of earthbound, staid fundamentals. (For some, this was still a matter of race. "The trouble with the ABA is that there are too many nigger boys in it now," Rupp, the Kentucky coach, supposedly once said.) In 1976 the ABA introduced the slam-dunk contest. It was, like most things in the ABA, "an act of desperation designed to get a few more fans to walk through the doors," Terry Pluto writes in *Loose Balls*, his history of the league. Erving ended up winning, with a high-flying takeoff from the free-throw line. *Sports Illustrated* would call the dunk contest "the best halftime invention since the rest room."

After that season, with ABA owners keen to shore up their finances, the league merged with the more established NBA. Beyond gaining new franchises, the older league, whose official logo, embroidered into all its jerseys, was the silhouette of a player in the very earthbound act of dribbling, was basically purchasing the rights to Erving—who had led the ABA in scoring for three seasons—and his modern, leaping style of play. Still, the NBA continued to have misgivings about promoting the dunk. It discontinued the slam-dunk contest, and, in these pre-Magic, pre-Michael years, the league couldn't shake a taint of the moribund. But on a November night in 1979 at Kansas City Municipal Auditorium, when I was about 10 months old, Darryl Dawkins changed all that. Thirty-eight seconds into the third quarter in a game against the Kansas City Kings, Dawkins drove two steps past his defender and dunked the ball so hard that the Plexiglas backboard shattered. Ever the showman, Dawkins, who played for the Philadelphia 76ers, told sportswriters after the game that he had been possessed by a "chocolate thunder." "I could feel it surging through my body, fighting to get out," he explained. "I had no control."

Several weeks later, Dawkins did it again, in a game against San Antonio, essentially crumpling the rim as he shattered the backboard—and suddenly he was a dunking sensation. Beat writers for the 76ers started a betting pool about when he would strike next. A stadium janitor in Detroit implored him to break a backboard so he and his team could show how fast they could replace it. The dunk got at something primal and maybe a little threatening to white refs and white NBA administrators. "When I dunked, no matter how nice and polite I'd do it, the refs would nail me with a technical foul for swinging on the rim,"* writes Dawkins, who, at one point became so popular that he had a column called "The Dunkateer" for a Philadelphia newspaper. "Even when I wasn't dunking, the refs were putting me in a straitjacket. Then the NBA summoned me to

---

* In a 2010 paper in the *Quarterly Journal of Economics*, researchers from Cornell and the University of Pennsylvania found that during the 13 seasons from 1991 through 2004, white referees called fouls at a greater rate against black players than against white players. The authors of the paper, titled "Racial Discrimination Among NBA Referees," concluded that the different rates at which fouls are called "is large enough that the probability of a team winning is noticeably affected by the racial composition of the refereeing crew assigned to the game."

New York and I was told that the next time I broke a backboard I'd be fined $5,000. Okay. That's cool. It's their fucking league after all. Right? Then a short time later I saw a commercial on TV: The NBA is exciting, blah, blah. Go out and see a game, blah, blah. Then there's a shot of me breaking a backboard. *What the fuck?* That was pure, unadulterated hypocritical bullshit."

Unadulterated hypocritical bullshit, but also a signal, finally, from the league that it was committed to the dunk. In a sense, it was following the lead of the college ranks, which had shrugged off the embarrassing dunking ban. The NCAA embraced characters like Dr. Dunkenstein, a.k.a. shooting guard Darrell Griffith, who led Louisville to the national championship in 1980, and the players of Phi Slamma Jamma, the irrepressible University of Houston squad of the early 1980s known for its prodigious jamming.

In 1984, David Stern, newly minted as NBA commissioner with a charge of expanding the league's appeal, reintroduced the slam-dunk contest, the one the NBA had unceremoniously retired when it had merged with the ABA. Two years later, Spud Webb won—still the shortest man to win the dunk contest. The embrace of the dunk is what would make basketball a truly international, glamorous sport: Michael Jordan famously won the competition in 1987 and 1988, besting Webb's teammate Dominique Wilkins, launching Jordan's global brand and a worldwide interest in the sport. Basketball officials, once keen to discourage dunking because of its expression of racial exuberance, now eagerly realized its commercial possibilities.

●

As a player, I'm no wild man. I'm more of the sensitive, spiritual, non-dunking sort, I guess you'd say. "Physical aggression, even camouflaged by athletic uniforms and official rules and intended to do no harm to Jews, was not a traditional source of pleasure in our community—advanced degrees were," Nathan Zuckerman, Philip Roth's stand-in, explains in *American Pastoral*. But from the hands of Swede Levov, the "household Apollo of the Weequahic Jews," the basketball shot carried the hopes of an entire community, writes Roth. "Through the Swede, the neighborhood entered into a fantasy about itself and about the world, the fantasy of sports fans everywhere: Almost like Gentiles (as they imagined Gentiles), our families could forget the way things actually work and make an athletic performance the repository of all their hopes. Primarily, they could forget the war." My own father was, like so many others in those postwar years, a would-be Swede. He had been born in Vienna in 1938, and by the time he was whisked away by his parents to America on February 14, 1939—exactly 40 years before my own birth— Austria had feverishly allowed itself to be absorbed into the Reich, and my father's infant passport was stamped with a swastika. "I was one of the last Jewish infants to be conceived in Vienna as it once had been," he wrote in his memoir, *Objects of Remembrance*. The family settled, eventually, in Cincinnati—the war, and murdered relatives, never forgotten. But for that new generation of Jewish-Americans, sports were an avenue of assimilation even as they were a way of consolidating ethnic pride. Who did not count Sandy Koufax as a hero, as a proxy for all that each of us could achieve? What Hebrew school didn't have a copy of *Great Jews in Sports*? (Thicker than you might imagine, by the way.) My father, however, who, along with my mother, bequeathed me so many things, including whatever natural physical potential—or lack thereof—I own, found that he was no Swede. He was thrown off the high school basketball team not because he was a terrible player—though he probably was—but because, clumsily, he injured too many of his teammates during practice.

It was my father, in the mid-1980s, who took me to the public-school playground a few blocks from our apartment and taught me the pick-and-roll. I grew up into a more competent athlete than he ever was, even if I, too, was never Swedian. I took it as a compliment, then, when, nearly halfway through my dunking year, an Austin friend invited me to join a team he was putting together for a spring basketball league. My chance, I thought, to show off my newfound jumping skills; to show myself, in my own mind at least, as deserving an entry in *Lesser Jews in Sports*.

It turned out we were a bunch of overmatched misfits. I knew we were in for a long season when a teammate, catching a ball just inside the free-throw line, proceeded to shoot the ball over the backboard. We had started the season 0–3. We were terrible. Our next game was against a 3–0 team. Our captain, always game, sent out an email to cheer us up and talk strategy. It also warned us not to get down if our opponents began dunking. I thought he was joking about the dunking. This was a league, after all, for punters and hacks and dreamers. Yes, we had lost three games, but nothing about those other teams was particularly impressive. We were the after-work, receding-hairline equivalent of Little League baseball, with just about every player doing all the little things wrong: the basketball version of allowing grounders to dribble between our feet, sitting cross-legged in the outfield, whiffing on the slow pitches.

Our opponents were the Ball Boys. I underestimated one of their players, a hefty man-child who spotted up and hit 3s all day long. In one of the more humiliating moments of the game, after they had opened up a 30-point lead on us, just as the man-child spotted up to shoot, two of my teammates flew in from opposite directions to try to block the ball; instead, they collided in midair, one of them sustaining a serious knee injury. As they crumpled to the floor, man-child coolly nailed the shot. And when he wasn't shooting, man-child passed the ball inside for an easy bucket.

I was more or less a nonfactor. Besides avoiding injury—I had put in too much time at the gym to see it all go the way of a badly turned ankle—my goal was basically just to rebound and play solid defense. At one point I found myself backtracking desperately as a guard pushed the ball upcourt. Just as we crossed the foul line, he threw up what seemed to me a wild shot. As I turned to collect the ball, I congratulated myself on thwarting what could have been an easy layup. Nice hustle, I thought. But just then, cutting inches away from my face: a pair of knees. One of the Ball Boys was in the process of catching the ball in mid-flight and dunking it. On me. I glanced embarrassedly toward the bleachers, where a collection of girlfriends sat, looking for a raised cell phone. I could have been said to have been "posterized." That is, captured in a humiliating position in a dunking poster to be taped up on a teenager's closet door. The Ball Boy hung on the rim just long enough for me to get out of his way before he dropped to the parquet. Something like a 50-point lead now for the Ball Boys. We ended up losing, I swear, by 71 points. In a 40-minute game.

I understood, suddenly, and from the wrong perspective, what Dr. J meant when he told the magazine *Black Sports*: "Dunking is a power game, a way of expressing dominance. It makes your opponent uptight and can shatter his confidence." By the time I got dunked on, I don't think we had much in the way of confidence to shatter. It didn't help that the same guy dunked three other times in our game, including a flying, two-handed, devastating jam, one so hard that the rim had to thwack itself back after the dunk. It reminded me of what Darryl Dawkins had to say about the curmudgeonly complaint lodged by Oscar Robertson, the splendid, high-scoring guard of the 1960s and 1970s. "A dunk is just two points," Robertson told Erving's biographer, Vincent Mallozzi. "I wouldn't spend three cents to go see a Slam Dunk Contest." Chocolate Thunder's response, one that serves as a rebuke to those generations of stuffy coaches and sportswriters who dismissed the dunk as nothing more than a gratuitous flourish: "Everybody says a dunk is only two points, but it gets your team hyped, gets the crowd all excited, and takes the starch out of other teams, especially when you dunk on somebody. And I always dunked on somebody."

7

Besting the Fates

Feb. 14, 192 days to go: *Today I turn 34. Carrot cake,*
for my birthday, made by Rebecca, with my favorite
cream cheese icing. "Come on, it's made of carrots,
it's practically a vegetable," she says. Two slivers
for me. Then, after Letterman, 45 hops on each leg in
the backyard.

The morning of my testicular surgery found me jacking off
in an act of filial duty. My mother worried about posterity.
Perhaps she imagined I would suddenly become a eunuch,
though I would still have one testicle left—this cancer spreads not
laterally, but vertically, to the brain. (That was the eventual explanation for the mysterious back pain: Doctors would soon discover
that the cancer had spread to my lymph system, with a second mass
lodged near my left kidney.) Maybe, she thought, my opportunity to
pass down my potential, my very DNA, was at risk of being snuffed
out. Having grown concerned about the future of the family line,
she suggested to Rebecca's father that we be urged to procreate. She
was only half-joking. So on the morning of my surgery, like a good
boy, I stopped by the Fairfax Cryobank, on the third floor of an

Austin office building, and nervously made a deposit. Sanctioned masturbation, especially in those circumstances, is not a sexy experience: As I went about fulfilling my obligation, my left nut swollen and sore, I had to keep an eye on the clock—I needed to be at the hospital by 10 a.m.

●

A few days earlier, an emergency room doctor, the one with the Daffy Duck tie, had told me I had testicular cancer. We left the hospital that clear and dry Saturday hand in hand with shaky steps. At Rebecca's parents' place we hunkered down to call my parents, my brother Josh (my other brother, Gabriel, is Orthodox, so it fell to my parents to inform him after Shabbos ended, at sundown), and my old school friend Nathaniel. When I got my father on the line, all I could manage to get out was that I had bad news. I started blubbering. "I'm so sorry," my dad said, so sweetly and sympathetically—and he didn't even know what the news was yet. I thrust the phone into Rebecca's hands. I don't know what my parents said to each other after we hung up, what tears they shed, and I still don't want to know. I myself shoved the news aside: That night, tickets already in hand, we headed with a couple of friends to San Antonio, to catch the Spurs play against who-can-remember-now. Watching the players, far below, flutter through the air, I lost myself in the game; Rebecca, not so much. The next morning, my testicular pain having subsided, I hit a tennis ball around with a colleague. I didn't let on about my news. Rebecca sat and watched. She was keeping vigil.

Monday found us in the waiting room of the Urology Team, in an anonymous medical complex in Northwest Austin, just off a road called Jollyville. Nearly all the patients were late-middle-aged

males, many of them, in the Texan style, with moustache and gut.
Posters along the walls advertised the da Vinci robotic prostate surgery; artificial sphincters, an antidote to urinary incontinence; and
penile implants. I whispered to Rebecca, who was now as fragile as
a dry, crackly leaf, that at least all I had was testicular cancer. She
wasn't in much of a joking mood.

Dick Chopp, who, I would later learn, had examined the under-
side of my own father-in-law and other male family friends in town
(I came to think of him as the go-to mechanic for men of a certain age), struck me as a very clean man. He was nearly bald, with
a shiny forehead. What was left of his hair was closely cropped. He
was immaculately shaven and wore the sort of glasses that barely
have a frame around the lenses. He had with him the ER doctor's
notes, including the results of the sonogram, but, being careful, he
asked me to drop my pants. With my pain at least temporarily gone,
it occurred to me that the emergency-room diagnosis had been a
mistake. Maybe the sonogram had lied, I thought, or maybe there
had been a misinterpretation. Chopp reached out, grasped my left
testicle, and nodded to himself. He was now the third stranger in as
many days to cup my bits, and he would be far from the last.

"You definitely have a tumor," he said, confidently but gently.
"We need to operate immediately."

"By the end of the week?" I asked.

"Tomorrow morning," he said.

The surgery is known as an orchiectomy, from the Greek word
orchis, or testis, and *ektome*, or excision. To my mind it sounded
like the bashful act of pinning an orchid on a prom date. I had pru-
dently decided not to ask about the logistics of the procedure before
it happened. It had to be done, so what was the point of learning the
details of my impending castration? Chopp had given me a choice
before the surgery: He could slip into my scrotum a neuticle, or a
prosthetic testicle. Evidently, some men, or maybe their partners,

want to maintain the two-testicle illusion. I waved the offer off—I found this option ridiculous and even, from what I had learned, dangerous—neuticles, made of silicone, can grow infected. But, still blessedly ignorant of exactly how my testis and I would be parted from each other, I had a question for him: Would I get local or general anesthesia? I mean, I asked innocently enough, would I be awake for the operation?

He shook his head. "You don't want to be awake for this surgery."

The four-inch diagonal scar in my lower abdomen is now faded, but that's where Chopp opened my guts and, in a medical sleight of hand, emerged victoriously with my left testis. I awoke shivery, woozy, and swaddled. The nurse wanted me to urinate, to make sure nothing had gone amiss. But as I stood over the toilet, swaying like a drunk, I wanted badly to avoid looking or touching *down there*. As I gingerly peeled away the dressing, I realized, even in my grogginess, that my groin had been shaven clean. I wondered to myself what poor schmo got that assignment—what first-year medical student found herself, as in some twisted initiation, having to take a Schick to the sack of some zonked-out dude. As it turned out, with chemotherapy soon to follow, my junk—what was left of it—would remain hairless for at least six more months.

●

My body back then was plotting an opposite trajectory from the one it was now charting, tearing itself apart instead of building itself up. As a sedentary cancer patient at an outpatient chemo ward, I soon endured a steady intravenous drip of medicines, from 9 a.m. to 4 p.m. daily. I was pumped full of water, too, to make sure the chemotherapy toxins didn't linger about my organs or bloodstream too long, for fear of lasting damage. I peed frequently, shuffling, stooped, to a bathroom with the burping, wheeling drip tethered to my black-and-blue forearm. By the end of each week, I felt a little more like an octogenarian and less like a 26-year-old. I took slow turns around the garden, my arm interlocked with Rebecca's as I hobbled along. In the shower, my hair fell out, sticking to the soap like metal flakes to a magnet. My face grew puffy from the steroids I took to blunt the side effects. I stopped shaving because I didn't have to—my beard stopped growing. Too weak to do much good, I ceased showing up at the newsroom. I grew weirdly sentimental, crying at scenes of death during episodes of *M*A*S*H* and the old movie *Goodbye, Mr. Chips*. Nauseated, vomiting on and off, I gather I was the closest I'll ever get to those first-trimester blues. I lost any real exuberance, emotional or physical, and if you had asked me just then if I might ever dunk, I would have thought you were being cruel.

8

A Dunk Contest

184 days to go: *12 pieces of chewing gum chewed, to keep myself from eating the food floating before me: Someone brought to the newsroom a box full of palmiers, crusty and golden and crunchy. I'm squatting 155 pounds. Back to 183 pounds on the scale. Is God punishing me for my birthday carrot cake? Or that meatball sub I ate last night: It wouldn't have been fair to the bread not to eat it along with the meatballs.*

Could Abe Lincoln dunk?

In the darkness of the theater, watching Daniel Day-Lewis play the president, most moviegoers, I'm sure, were captivated by the vote-getting needed to abolish slavery. But I began wondering whether Lincoln could have dunked, had basketball been invented when he was a young man. As far as I know, none of the many Lincoln biographies speculates on the matter. On the one hand, he was 6'4" and grew up a rail-splitter. He must have been a strong man. And you get the sense—maybe it's just because of the Lincoln Memorial—that the guy had big hands, big enough to palm a basketball. On the other hand, he doesn't look like the dunking type: thin of bone, and narrow-kneed. Preoccupied, certainly. And did they even jump back then? I mean, beyond Lincoln's seriousness and studiousness and presumed disinclination to jump; I mean, why would they have jumped after childhood? They were not playing pickup baseball or basketball, after all. Did adults even play sports before the late nineteenth century?

These were the sorts of thoughts that occupied me in the early spring. I was in Houston for the NBA's All-Star weekend. To be more precise, as most fans gathered in Houston's Toyota Center, a sleek decade-old stadium large enough to seat more than 18,000 people, I found myself across town in the 25-year-old George Brown Convention Center, attending the Developmental League All-Star festivities. The D-League is basically a minor league for the NBA. Drafted players too green to play for the mother ship end up in towns like Bakersfield, California; Fort Wayne, Indiana; or Canton, Ohio, where they stay warm for the big time. Houston had dressed up the convention center nicely. Outside the heavy black curtains that separated off the makeshift 5,000-seat arena, fans crowded into a carnival of cotton candy, basketball relics (one display included Michael Jordan's University of North Carolina sneakers), and amusements meant to show off the game. I had dropped by for the Boost Mobile Dream Factory Dunk Contest: I figured even if the players didn't have quite the skills to play in the NBA, they were still just as athletic. Told from a young age that they might have NBA-level talent, they were on their own quests, enduring yeoman salaries and coach-class travel for that moment in which, they trusted, their true selves would be spectacularly revealed. Being named a D-League All-Star struck me as a kind of backhanded recognition; most of these players would have gladly traded this spotlight for the anonymity of the NBA bench.

The dunk contest is, in many ways, a silly affair. Divorced from a game situation, it does not have to abide by the rules of basketball, such as dribbling. But that makes it a freely athletic event, one that puts emphasis on creativity. "I developed the dunk shot because it was a challenging thing to learn to do," Julius Erving, one of the great in-game and out-of-game dunkers, once told *Black Sports*. "After you learn to do it one way, it's a challenge to learn to do it another way and right on up. There's an infinite number of dunk shots which I don't possess. And there are psychological considerations, too. I wouldn't like to start feeling that I've done all that I could do and the coming years would be a matter of repeating what I've already done."

No one left her seat at halftime of the D-League All-Star game. If anything, the small arena grew more crowded: It was time for the dunk contest. In each round, each of the half-dozen participants had a minute in which to complete a dunk. Despite the buzz, the dunkers themselves appeared lackadaisical. When they missed, they slowly gathered the ball and repositioned themselves. They didn't show any urgency, and I couldn't tell whether they didn't care, or didn't want to seem to care. Or maybe they were just regathering themselves after an explosion of effort. The emcee, a young, upbeat, blue-jeans-wearing guy with slicked-back hair, was growing increasingly urgent and disbelieving with each round ("Fifteen seconds left!") as the dunkers appeared increasingly nonchalant. One, Dar Tucker, of the Reno Bighorns, moved with the opposite of purpose. He kept slowly retying his shorts, as if no one was there.

The dunks were, of course, amazing: One guy dunked a ball tossed by another player from up in the stands; another dunked over a seven-foot teammate; in a third dunk, the player stood away from and behind the basket, lobbed the ball in the air, ran, picked it up off the bounce, and windmilled it in. "Houston, make some noise!" the emcee kept shouting. Yeats once summed up youth as a series of "moments of glad grace," and I felt I was witnessing one such moment after another.

And then, glancing to my right, only several feet away, I noticed a very familiar looking black man, lanky, handsome, older, cap pulled down, leaning against the very same banister as I was. He had with him a retinue: a couple of grandkids, maybe, one a seven-year-old with a mini-Afro, a wife, a daughter, a brother or friend. Only a few weeks earlier, at the presidential inauguration, George Stephanopoulos had confused Bill Russell for Morgan Freeman, and I didn't want to be that guy. But my mind was buzzing with recognition, racing through the possibilities. Not Denzel Washington, I was sure of it. And then I got it: "Is that Dr. J?" I whispered to an usher next to me. He grinned and nodded. I leaned past the usher, extended my hand, and told Dr. J I didn't want to bother him but that I was a big fan. "Thanks," he said.

Ugh. Of course I wanted to bother him. I walked just past him. And then I stopped. My knees weakened. Maybe it's just that it had been a long day—I was dehydrated and tired—but I started getting nervous and tingly. After all, here was Julius Erving himself, and I couldn't let this opportunity slip by. Could he lend me special advice, some heretofore-overlooked bit of know-how, that would help me dunk? But there he was, with his family, enjoying himself, a guy who a hundred thousand times or more is asked about the myriad mysteries of basketball. I lingered, trying desperately to figure out how, actually, I could bother him. An arena camera crew came over, and suddenly the emcee said: "We have a special guest in the stadium tonight: Dr. J himself!" And the bright light above the camera flipped on and Dr. J appeared on the jumbo television floating above the court, his cap doffed as he acknowledged the crowd, which offered, I swear to God, the biggest cheer of the night. I looked up on the Jumbotron, and there I was, just behind Dr. J, looking up at said Jumbotron. Of course, I wasn't paying any attention anymore to the dunks on the court. Little balls of sweat were forming at the edge of my scalp. With each second of lingering I felt like it was getting more and more obvious to Dr. J and his group that I was floating just behind, weirdly. I had become a stalker of dunkers.

One of the dunkers, for his final dunk, leapt over a table. The camera crew had moved on, and, desperate to spark a conversation, I half-shouted to Dr. J whether he approved of props. He glanced up from his iPhone: "No, I'm not so crazy about them," he said. Okay, I thought, I've made a connection! But I was frozen. The whole thing was so asymmetrical: I wanted desperately, sincerely, to chat with him, and for all he knew I was just another dude wanting a photo.

Then Julius Erving and his family shifted toward the exit, and I knew I had to make my move. I sidled up to him: "Dr. J," I said, "I'm actually a journalist working on a book about the history and science of jumping . . ." "Jumping, huh? Sounds interesting." And then, something he must say daily, something that gives hangers-on just enough hope: "Get in touch with me at Doctor J Enterprises dot com, and we'll see what we can do."

I turned my attention back to the dunk contest, which was wrapping up. The cosmic weirdness of my afternoon was suddenly compounded when I realized that Darryl Dawkins, Chocolate Thunder himself, was a judge. I actually had a copy of his book, from the library, at home, and I was certain that I was one of the very few people in the world who had bothered to borrow it. He looked like a bigger Michael Jordan, if such a thing can be said. A massive shaved bald head and enormous, wide shoulders. He was stylish, too. He wore, yes, chocolate-brown pants and a blazer that looked like os-trich leather but, on closer inspection, was a kind of cotton with a meticulous tie-dye pattern. Chastened by my weak-ass effort with Dr. J, I tailed Dawkins after the dunk contest. And then I stood meekly aside as a 19-year-old posed with him for a picture, his hand resting around her waist, not all that far from her hip. I wondered what his wife and adult daughter, both done up, made of the scene as another young woman came up to him. Maybe it was half BS, but

I remembered that much of his book was about the ease with which he had bedded women. "No matter how pretty or sweet-smelling your wife (or your girlfriend back home) was, there was always a girl on the road who was prettier, smelled sweeter, had that certain walk that made her ass pop, and knew how to come at you," he wrote of his time as a pro.

Still, in front of me in Houston years after his career had ended, he appeared to be a genuinely down-to-earth, earnest family man. He smiled and patiently had his picture taken with little kids. And then, when the crowd finally subsided, he sat with his wife and daughter as the D-Leaguers played the second half of their all-star game. He picked an aisle seat, and I wondered whether it was a self-defense mechanism, to prevent punters from crowding him. That's where I came in. I crouched by his seat and told him about my determination to dunk as the announcer kept shouting the score over the PA system. He smiled, but appeared completely uninterested—and unmoved that I had actually read his book. I had the same sort of feeling that I had had with Dr. J: I'm earnest, I'm serious, I need to talk to you; and these guys, fair enough, had other things to do. He told me to contact him through the Brooklyn Nets. Odd, in a way: Here I was, face to face with the man in some convention center in Houston, and I'd have to call up Brooklyn to get in touch once he was back in Georgia. (Neither Dr. J nor Darryl Dawkins returned my messages.)

The dunk competition over, I sidled up to winner Tony Mitchell of the Fort Wayne Mad Ants: okay, a real-life dunker, one who had just won the slam-dunk contest, one I would not be cowed by, and not a big enough celebrity not to talk to me. He had just been given mini–rubber bouncy balls—all the All-Stars had—to throw into the stands. In the shadow of the rim, I asked him if it felt peculiar to dunk in an artificial situation in front of several thousand people and not in the rhythm of a game. To me it seemed like the difference between gymnastics and rhythmic gymnastics. I'm a tall guy, but I strained to hear him: "I'm just trying to be a role model, to put on a good show." I looked up at him. It seemed to me a non sequitur. "Did jumping so high come naturally to you, or was it something you had to work at?" "It's all about having a good time," he said, and he was kind of looking past me, into nothingness. I suddenly understood why sportswriters, beaten down by years of answers like these, start asking clichéd questions. As with so many other things that afternoon, I sensed a separation between words and action, between my world and the one to which I aspired. I was trying desperately to work at dunking, to articulate, in my own mind and body, all the things I would need to do to execute it; like studio musicians who can sight-read and perform a fresh composition, these dunkers need only tell themselves whether they would go with the tomahawk or the windmill jam. "Years of athletic training teach this; the necessity of relinquishing doubt and ambiguity and self-inquiry in favor of a pleasant, self-championing one-dimensionality which has instant rewards in sports," Frank Bascombe, the eponymous narrator of Richard Ford's novel *The Sportswriter*, tells us. The implication, of course, is that there is a wide chasm between the crippling deliberation a writer faces and the natural ease of athletes. Considering this difference, I remembered what Dr. J had once said, that as a kid he had felt "these different things within me, certain moves, ways to dunk." The line imagines talent as a kind of natural underground reserve, restlessly awaiting a fissure—an opportunity to escape to the surface. And for the rest of us, the ones who do not feel an innate bubbling-up of talent? I'm not sure what Dr. J would have been able to tell me, other than you can't drill your way to oil if there's none down there to be found.

There's a saying that you can't teach height, and standing there, peering up at Tony Mitchell, I understood its truth, that there's something incommunicable about talent. So I left him, making my way off the hardwood, beneath the basket, through the maze of the NBA-related carnival and the throb of amped-up hip-hop and into the Houston sunshine, pleased to make my way back to Austin.

180 days left. *Just past halfway! I am the proud owner of a two-pound jug of Dymatize's "Elite Gourmet" strawberries & cream-flavored powder, the "perfect blend of Ion-Exchange Whey Protein Concentrates, Whey Protein Isolates, and Whey Peptides, combined with Milk Protein rich in natural micellar caseines & caseinates plus added calcium caseinates." Translation: I'm going to get HUGE! To max out my muscle-building protein (you find yourself using phrases like "max out," "push it," and, admiringly, to the mirror, "You the man," when you lift enough weights), the gym people advise me to down these protein shakes. So I've decided to greedily chug Dymatize, mixed with nonfat milk, before and after my workouts. It tastes like strawberries and cream just like Yoohoo! tastes like chocolate.*

Progress report: The good news: by early spring, feeling yet stronger, I could easily jam a tennis ball on the rim at the basketball court down the street from my house, at a Boys & Girls Club. The "Dream Court," according to the circular logo set into the floor at midcourt: "Be Great" was stamped on each backboard—and I felt mighty good. Even as I had put on muscle, I had slimmed down, dropping two pants sizes. I looked fit. A thick blue artery made itself visible, running the length of my left biceps, and Rebecca had made impressed noises about my abdomen. I had eliminated alcohol from my diet, not that I was ever a lush, and there was an overall sobriety to my approach: monklike, purposeful. I wasn't all that much fun that year. Rebecca sportingly—and helpfully, as chief chef— joined me in these salad-heavy days. (Maybe she was just abiding by her oath: "In sickness and in health, in lean times and in fat.") We avoided going out with friends, partly because we didn't want to be led into temptation. My cheeks grew drawn, baby fat gone. I took a pleasure in the asceticism—in being trim and strong, but also in the mission of it. Even in front of the television, I'd try to palm the ball. I found myself alert to the point of restlessness. I needed to take advantage of every day I had left—now six months to go in my project. I had no "cheat days." I didn't go to the gym on the weekends, but on Saturdays I played soccer and on Sundays I played basketball. Diet-wise, I was always disciplined. I grew to have a distaste, if not quite a revulsion, for sweets. I more or less had no dessert. It was like I was trying a new religion. Something abstemious and, on the face of it, unhappy. But I had become a devotee. I had long been a shambling, distracted type, the social sort that would cry out on forest hikes at everything worth remarking on, needing my enthusiasm confirmed by another, and now I felt closer to the rigidity and quiet, solitary pride of purpose of a Marine. I was sipping more from the cup of life, even if it was filled only with water. Sure, I was tired, but also bright-eyed. "I didn't know you could get that high," a teammate of mine shouted after I leapt up and grabbed a seemingly overthrown basketball. I didn't know I could, either.

And here I was, dunking an object—even if it wasn't exactly as large as a basketball. Size matters because the larger the diameter of the ball, the higher you have to hoist it over the rim. But working my way up—a strategy recommended by Todd Wright, the Longhorns strength coach—had the advantage of improving my jumping tech-nique: the smaller the ball, the easier it is to squeeze, and the more confidently you can swing your arms to gain a bigger leaping mo-tion. Your arms' swing speed, in turn, naturally dictates the cadence of your foot turnover; the faster the armswing, the quicker the ca-dence, propelling you faster and, thus, higher. Partly this is because your arms counterbalance your legs, allowing you to speed forward without toppling over; partly it's because the upward–downward piston motion of your arms as you run adds to the force and quickness with which your feet strike the ground and leap up again. (For these reasons, it's tough to run for a bus while carrying shopping bags or shouldering a purse because you can't really chop your arms up and down. You can get a lot more speed if, like a schoolboy, you're wearing a backpack.)

I had a thrill of exultation as the ball, having been pushed through the rim, bounced up satisfyingly from the pavement. Excited by my progress, but wanting to be scientific, I got my stepladder and tape measure from home and brought them back to the court. I climbed to the top of the ladder and, face-to-face with the "Be Great" stenciled on the backboard, I held the rim for balance as I dropped the tape measure down like a pendulum. It swung briefly before settling its weight on the ground. I peered at the result: The rim was 9'10" off the ground—two inches short. I blinked with disappointment. I stood there on the top step and rolled my eyes up to the sky, cursing the universe, cursing the Boys & Girls Club and its false dreams, and cursing myself. The goalposts had moved on me. I came home, shoulders hunched, and slammed the front door behind me. Our house, an old Craftsman bungalow always restless on its foundation, shook. "Goddammit," I shouted to a surprised Rebecca. "The fucking rim is fucking short."

I wasn't the only one finding that my self-worth and identity were intimately tied to dunking ability. In March, a few weeks after I went to the dunk competition, news came out of Chicago that Derrick Rose, the stony-faced point guard of the Bulls, had been finally cleared to play by team doctors. Eleven months earlier, in the 2012 playoffs, Rose, known for his springiness and hard-driving ways on the court, had torn the anterior cruciate ligament in his left knee during a relatively simple maneuver known as a jump-step, one he had probably executed a hundred thousand times in his career. Despite the medical clearance, however, Rose decided to remain on the Bulls' bench, wearing an elegant series of suits, through the end of the season, in May. He frustrated Chicago fans and talk-radio-show hosts, who said they couldn't understand his unwillingness to help his team. But Rose had his own medical standard, reporters learned. He had told the Bulls that he would not suit up to play until he could confidently dunk off his left foot "in his mind."

About the same time, Brooklyn Nets guard Deron Williams, a 6'3" All-Star, was dealing with a much smaller medical malady of his own. His ankles wobbly, doctors prescribed plasma therapy, a treatment in which his own blood was spun in a centrifuge and then injected into the injured area. A week later, in February, Williams made an improbable confession to journalists. "I can't dunk," Williams said. "I can't jump. Even if I tried, off one leg I can't dunk. I can dunk off two, but if I just tried to dunk off my left leg, I can't dunk." The writer suffering from a creative block; the aging mathematician losing her facility for numbers; the ballplayer robbed of hops—talent can depart softly or in a snap.

But it can return, too, and for Williams it did. Following an April playoff game against the Rose-less Bulls in which he performed a reverse dunk, he updated reporters asking about his health. "I think the dunk showed it," Williams said. "My legs feel good, my ankles feel good right now. I'm excited."

That's how I told myself to feel—I was jumping higher than ever, after all. I admit to the pleasure of achievement. I had managed something quite real, something measurably better than anything I had done before. And something to build upon. I was a certified dunker. Of sorts. Even if it was just a tennis ball and even if the rim was just a wee bit short.

9

Girl on Fire

166 days to go: *After two weeks of consuming daily servings of Dymatize's "Elite Gourmet" strawberries & cream-flavored powder—"a flavor and texture profile beyond belief"—I learn of an unfortunate, known side effect: The powder that brags of "instant mixability" doesn't mix so well with my gastrointestinal tract. So out with the whey protein powder. It's just not fair to Rebecca. Or to me, really. And, in my effort to further ratchet down my carbs, I'm officially resigning as an eater of oatmeal. Just grapefruit and maybe a sausage—for protein—for breakfast. And a glass of nonfat milk. And a small serving—make that two servings—of nonfat yogurt. With a teaspoon of fig jam mixed in. An avocado and tomato salad for lunch. Small helpings of hummus throughout the day. And bits of fruit, of course. Dinner of pork loins in a smoky Mexican salsa machaca. Squatting 165. Using the 20-pound dumbbells for the lunges. Four sets of four pull-ups. I'm comin' for you, rim!*

The ride to Waco is a straight shot up I-35, past the Vietnamese and South Asian neighborhoods of north Austin, past the cookie-cutter suburbs of Round Rock and Pflugerville (whose newspaper is the Pflugerville *Pflag*, and whose festivals include—I kid you not—Spring Pfling, Deutschen Pfest, and the Pfall ChiliPfest; you can imagine how they spell the name of their weekly farmer's market), past the Ikea, past the pickup truck dealerships and then the limestone quarries. After the town of Temple, known for its VA hospital, and not far from the major Army installation of Fort Hood, the road flattens and the landscape empties out into a quilt of cotton fields.

About 75 miles out of Austin, one gets to McLennan County, a notch in the Bible Belt. Fire-and-brimstone territory. Rebecca and I headed north in early spring, and as we crossed the county line we found ourselves passing enormous stacks of timber that had been set afire, like funeral pyres for giants, sending columns of smoke twisting skyward. Surely it was just a range-clearing burn, but the effect was suitably Old Testament. Our destination was Waco, home to Baylor, a Baptist university so committed in its beliefs that models in the art school remain clothed. The school's sexual-misconduct policy includes "homosexual acts"—alongside sexual harassment, incest, and adultery—as "misuses of God's gifts." Its president is Kenneth Starr, renowned as the special prosecutor who investigated the sexual proclivities of Bill Clinton. What had put Waco on the map for most Americans were the Branch Davidians, the messianic religious group that had holed up in a compound just outside town with an arsenal hefty enough to kill four federal agents and hold others off for nearly two months during a 1993 siege—before 76 of the Davidians were killed in a fiery final showdown.

Beyond a downtown that includes a handful of mid-century office buildings, relics of a time when cotton was booming, the town is a quiet, steamy, low-slung place, home to feed stores and farm equipment dealerships. You can find the Dr Pepper Museum here; above it, on a rarely visited floor, is the Beverage Executive Hall of Fame, an homage to what American capitalism can do with the wonders of carbonation.

We had come to see Brittney Griner, the six-foot-eight All-American whose Lady Bears were scheduled to play the Lady Long-horns of the University of Texas. The previous season, Griner had led Baylor to a record of 40 wins and zero losses and the national championship. This year was shaping up as no less successful, and if the Baylor team won tonight they would clinch the Big 12 conference championship. I didn't care: I wanted to see her dunk. Partly I thought Griner might act as an odd kind of mirror to my own efforts, a vision in what the unexpected looks like. That would be its own inspiration, even if Griner's dunks looked on television like the most natural thing in the world. Partly I felt that seeing this woman dunk would offer refreshment, by simple virtue of stripping the layers and layers of machismo the dunk had acquired in our culture. But, I would learn, a woman's dunk grows plenty of layers of its own.

She had dunked 13 times in her college career, more than any other woman in the history of competitive basketball. I had read that her hands are nine inches long, from the base of the hand to the tip of the middle finger, and nine and a half inches wide, allowing her to palm the ball easily. (My hand is only five inches long by five and a half inches wide; cake-decorating hands, a friend calls them.) I also had learned that she wears size-17 shoes and has a wingspan of 7'3½". I had expected, then, to watch an unwieldy Goliath who more or less artlessly manhandled competition. During her sophomore year, after all, Griner was suspended for two games for throwing a punch at a Texas Tech player, breaking the girl's nose.

We were down on the floor during warm-ups, mingling with the college band and the press corps and the VIPs and the contestants called upon for the in-game breaks—the tricycle racers and the baton twirlers. Coolers of Dr Pepper were everywhere. There were a lot of very tall women—it was homecoming for the Baylor Lady Bears—but Griner, taking sleepy warm-up jump shots, was unmissable. It was the way she moved: She had that athlete's way of efficiency masquerading as laziness; a kind of ease composed of no unnecessary motion. Griner smoothly caught the ball; sprung up; released. That was it. She had a long Modigliani face and dreadlocks dropping down her back. She was ever so slightly pigeon-toed and small-stepped, the way many athletes are. I'm slightly splayfooted with big, clunky steps: If I manage to dunk, I thought, it will be the most strenuous thing I have ever done.

With the game about to start, the teams retreated to their locker rooms. No dunks yet from Griner. We made our way to our seats just before the introductions. A crew of 50 elementary-schoolers performed a "jumping jamboree," hopping on pogo sticks while jumping rope in a series of exquisite routines. There was something spectacularly un-self-conscious about them, even as they performed in front of the amped-up crowd. They were boys and girls, big and small, chubby and stick-thin, and all of them very capable jumpers. It made you think that we're each born great, and then somewhere along the line we fritter away our greatness.

The lights went down, the smoke machine whirred, and over-head on the Jumbotron appeared a highly produced video of Kim Mulkey, the coach of Baylor, addressing her attentive players. "Con-gratulations on your championship," she said. "Now it's time to get back to business." As she continued in this vein, the video cut to shots of the players pumping iron, or, one, bespectacled, in the class-room, taking notes. The game was a sellout, and the crowd stood now—old men and young women, boys, mothers, college students, virtually all dressed in dark Baylor green—and stomped their feet, making the arena feel like the inside of a bass drum. Some of them pawed the air as if they were restless bears. I felt a shiver and Re-

becca's eyes grew moist: "So much being made of a women's athletic event," she said. She had grown up going to women's basketball
games at the Drum, the arena at the University of Texas. Rebecca
was taken there by her father, who had long preferred women's ball
precisely because they don't jump—not like the men. He is a man
fond of telling stories about growing up in Middletown, in upstate
New York, in the late 1950s, stories that invariably include both the
first and last names of their characters ("Harry Golumboski"), and
for him the Lady Longhorns play something closer to the grounded,
fundamental game of his youth. Or most of them do. The smoke
took over the floor as the announcer, calling out the players, reached
"BRITTNEY GRINER!" and the crowd, still on its feet, went ber-
serk. Rebecca was torn in her rooting interests: "I don't want to be
disloyal because Daddy's a Lady Longhorns fan!" she shouted in my
ear as she clapped in spite of herself. A women's choir from a Baptist
church sang the national anthem and it was finally game time.

Griner had had a relatively unproductive outing the last time
she'd played UT, two weeks earlier, scoring only 14 points. But
she had recorded a dunk, and the next day's newspaper made sure
to mention that near the top of the story. No wonder: Only seven
women have ever dunked in the history of college basketball. The
simplest explanation is that women just tend to be shorter and don't
jump as high. Female college basketball players have an average vertical leap of 19 inches—just about my vertical—and the male players
have one of about 28 inches. Similarly, researchers at the University of Missouri studying medical students in their 20s and their
spouses found that the average guy outjumped 95 percent of the
women. The reasons, according to a range of scientific papers, have
their roots in muscle composition and hormones. When females
hit puberty, their takeoff force actually decreases, just as the boys'
increases. Puberty, not coincidentally, is when males begin pro-
ducing more testosterone, a naturally occurring steroid that helps

build muscle mass. The testosterone also increases the size of motor neurons, which recruit muscle fibers to contract. Henneman's size principle, named for a 1960s researcher who examined triceps contractions in cats, holds that to move a load—such as propelling your body into the air—motor units (which make up motor neurons and their associated fibers) are recruited from smallest to largest. Practically speaking, that means smaller slow-twitch, fatigue-resistant muscles, ones used for walking or jogging instead of jumping or sprinting, are activated before their fast-twitch counterparts. This makes sense: Walk before you run, the saying goes. Scientists call this recruiting of muscles in the exact order of the capacity to exert force "task-appropriate" recruitment. Think of it as moving through the gears of a car as you accelerate. The chief benefit of this internal gear-shifting is that it minimizes exhaustion—or engine wear, to stick with the car metaphor—by using the fatigue-resistant muscle fibers first. For some of us, our ratio of muscle composition heavily favors fatigue-resistant, slow-twitch muscles over the fast-twitch ones; we can't truly knock it into fifth gear. Slow and steady wins the race, we're told, but that doesn't mean you can dunk. Usain Bolt or Brittney Griner, meanwhile, because of some mix of hormones and genetics, have a muscle ratio weighted toward fast-twitch muscles.

But some doctors and researchers offer a competing explanation for the difference in male and female jumping ability. While individual muscle fibers are larger in men than in women, female muscle tissue does not differ from a male's in its initial potential for force development. In other words—and this is a theory Rebecca ascribes to—perhaps females are so socialized from birth, in a thousand and one ways, including, most obviously, being dressed up in pink and being told they're adorable (as opposed to being complimented on, say, their strength), that any inclination to jump high is wrung out of them. It's a theory almost impossible to test, since a newborn girl would pretty much have to be raised in a cave to be inoculated from societal pressures.

Griner was no cave dweller. She grew up in Houston, and by the time she was in high school she was a standout player. In her senior year, she dunked 52 times in 32 games, including seven dunks in one game. In another game, she recorded 25 blocked shots. In short, she was playing an above-the-rim game while everyone else was below the rim. The afternoon we were in Waco, Griner methodically took over the matchup with the Lady Longhorns; by halftime she had recorded a double-double—20 points and 10 rebounds. For all its potential power, her game struck me as delicate. Lots of little drop-steps and then turnaround jumpers, with quick half shots only a few feet from the rim. She reminded me of Kareem Abdul-Jabbar. So much touch. Even Texas's own giantess, a freshman named Imani Stafford, who is 6'7", appeared to have a wingspan too short to contest Griner's shotmaking. "Brittney does what Brittney does," Stafford said after the game. "She definitely has a big body. You think you have an open shot and you don't."

The student band shouted "Dunk it!" just about every time Griner got the ball, but she was often triple-teamed by a storm of Lady Longhorns, their arms flapping like rain against a cross-country semi-truck. Then she would pass to an open teammate for an easy bucket or quickly spin away from the extra defenders and scoop the ball against the glass and into the hoop. I thought this must be the treatment that the earliest male dunkers got: the double-teaming, the shouts of encouragement from the crowd, the sense of a player distinct from the others.

Baylor clinched the Big 12 Championship in that game. At the final buzzer, white, gold, and green confetti blew everywhere. After posing with her team with a trophy, a huge, toothy Julia Roberts smile spread across her face, Griner ran and dove across the confetti-strewn gym floor, as if it was a backyard Slip 'n' Slide. She lay still for a second, absorbing the feel of the vast buzzing space spreading over and around her—a space that, in that moment, was all hers—and then, before getting up, silly, giddy, she made snow angels. The press gaggle crowded about her, taking photos as she acted her age, a kid having fun. The Alicia Keys song "This Girl Is on Fire" blasted over the PA system, and there was little doubt to whom it was dedicated. ("She's just a girl and she's on fire . . . Filled with catastrophe, but she knows she can fly away.")

I had wanted to ask Griner, in the media room afterward, whether she would have rather taken the lighter performance with the dunk that she had turned in two weeks earlier, or the over-whelming one without the dunk. Or, I thought I might ask her what she made of the comments by Geno Auriemma, the coach of the University of Connecticut women's team, that the hoop ought to be lowered to nine feet to promote dunking and excitement in the women's game. But Griner was a no-show. "She's probably far from the arena now," her coach said. "She must be hungry and went to get something to eat." A sports reporter covered his mouth with a note-book and whispered to me that Griner's guarded around the media. "She knows people say some cruel things about her." "What kind of things?" I mouthed back. I had no idea what he was talking about. "Her masculinity," he said.

Many of us are unsettled, too often unattractively so, when someone operates outside the boxes in which we expect them to perform. Back at home, it wasn't hard to find mean-spirited com-ments about Griner tucked beneath online videos of her highlights. In an odd, unkind way, the very attributes of her game that are cher-ished among young males—foremost among them, the ability to dunk—appeared to be held against Griner because she is female. It's an ancient taunt leveled at assertive ladies: "Women are soft, mild, pitiful, and flexible; / Thou stern, obdurate, flinty, rough, remorseless," Richard, a would-be king, sneers at Margaret, who has taken up arms to defend her husband's throne in Shakespeare's *Henry VI, Part 3*. Griner was a latter-day virago, distinguished because of her heroic, "manly" abilities and demonized for the same qualities.

"This is someone's child," Mulkey had said in one press conference in response to a question about the derogatory anonymous comments posted about Griner, ones that accused her of being, in a word, unwomanly. "This is a human being. She didn't wake up and say 'Make me look like this, make me six-foot-eight and have the ability to dunk.' This child is as precious as they come." Academics and critics across the nation weighed in on what the sub rosa discussion of Griner's gender meant about women's sports and the American male gaze. Because Griner lacked an "erotic dimension" that would appeal to straight men, "she has found limited use within the national imagination," David Leonard, a cultural and gender studies professor, wrote in *Slam* magazine. Griner, for her part, seemed to shrug off the insults. "I'll go and search my name on Twitter, to see some of the things they say," she once told reporters. "[School officials] tell me I shouldn't read the blogs: They're mean. But it doesn't bother me."

For me, trying to dunk was shaping up as an effort to overcome my natural physical shortcomings. But for Brittney Griner, the dunk was an exuberant expression of that innate physical identity that was being picked apart by a horde of anonymous bloggers, sports commentators, and academics. In a man's world, the dunk is a flourish on a basket worth two points like any other—but each time Brittney Griner dunked, the act shouldered all kinds of interpretations. Were her jams feminist statements, each forceful slam another strike against that glass ceiling, each leap a vault into the rarefied air normally occupied by men? Or were they simply outpourings of awesomeness? The dissection of Griner's dunking habits seemed itself unfair: Why does it have to be so complicated for Brittney Griner to dunk a basketball when it's pretty much just the opposite for a guy who dunks? The probing of her dunking domination was emblematic of the greater burden that Brittney Griner, through no fault of her own, was forced to bear. It's a situation Rebecca, and probably a lot of other non-dunking women, can relate to. Rebecca doesn't like playing football, but one year she came along, out of duty, to an annual ragtag Thanksgiving two-hand-touch game I organize. She turned out to be the only woman to show up, and besides, the numbers were odd. It would have been easy for her to sit and watch. But even before the game actually began she felt it was too late to back out: As the lone woman, she didn't want to allow anyone to think she wasn't game just because she's female. She didn't want to let the side down, as it were, and the thought of sitting in the bleachers cheering on the boys didn't suit her image of the world. On the other hand, she sucks at football—she'd readily admit this—and all she would accomplish by playing would be to look and feel pretty hopeless. It's the kind of no-win calculation that only women need to make.

Late that Saturday afternoon in Waco, in the pressroom after the Lady Bears had clinched the conference championship, Kim Mulkey, pert-nosed, with short, peroxided hair, was asked about how she gets her team prepared as they drive toward another national-championship run. She still wore the enormous heels on which she had squatted during the game; they themselves, absurdly impractical for coaching on the hardwood, seemed to me a concession to the demands that women maintain some trapping of conventional femininity even in the context of a conventionally masculine pursuit.

"I told them to dig deep into their souls and find ways to keep winning," Mulkey said, in a very Waco answer, one that somehow combined basketball with spirituality. She was pastor as much as coach. "The memories are all they will have in life. The trophies are collecting dust, but nobody will take those memories away. We're blessed to have kids that understand that."

●

The Baylor team closed out their regular season with several more sparkling victories. In Griner's final home game, in the second round of the season-ending NCAA tournament, she tweeted from the locker room at halftime that she would dunk twice more; she then delivered. The team was favored to repeat as national champions. But only four days later, in the Sweet Sixteen, a scrappy Louisville team disrupted Baylor's rhythm, ganging up on Griner. She didn't score in the first half, and her teammates failed to drain open shots when she got them the ball. In a major upset, Baylor lost 82–81. Three weeks later, Griner, already out to her friends and family, declared publicly that she's gay. She said Mulkey had told her players not to be open about their sexuality because it might hurt recruiting and the program. "It was more of an unwritten law [not to discuss your sexuality]," she told ESPN. "It was just kind of, like, one of those things, you know, just don't do it. [The coaches] kind of tried to make it, like, 'Why put your business out on the street like that?'" In a prepared statement, Mulkey told reporters that she could not "comment on personal matters surrounding any of our student-athletes, but I can tell you Brittney will always be a celebrated member of the Baylor family."

Turning professional, Griner was the first overall pick in the late-spring WNBA draft by the Phoenix Mercury. In her first WNBA game, unshackled by the name-calling and sexuality-suppression that lingered in Waco, she dunked twice.

10

"Dayenu"

163 days to go: *Foggy, frigid spring morning, River-side Park in New York City. On the road, so no squat-lifting. Another kind of misery: 90 squat jumps. 120 jumps up and down from a park bench. 600 ankle hops. And I can just get my butt below my knees when I squat flat-footed! My right knee is bothering me a bit. At night, after everyone is asleep: three spoon-fuls of Häagen-Dazs dulce de leche ice cream from my parents' freezer. I pledge to eat an entire pint of that stuff when I'm done with my dunking year. Still at 182 pounds. My mom thinks I look so very handsome.*

In March, a shade past the halfway mark of my dunking proj-ect, I headed to New York to see family, friends, and, of course, Polly and Jamie. At one lunch at an upscale French restaurant off Park Avenue I found myself doing battle with my stomach: A baguette lay across the table, daring me. As the waiter cleared our plates at the end of the meal, he asked if we wanted any tea, coffee, or cookies. "Yes, cookies," one of my dining companions, familiar with my project, told him. "But"—he gestured with his head toward me—

"he won't have one." I felt like a 15-year-old whose parents won't let him take even a sip of wine. Or maybe a prizefighter accompanied by his manager.

My friend was right, of course. The next day I was scheduled to meet with Polly and Jamie to see what, if any, progress I had made. I knew I was slimmer, I knew I had done a ton of squats, and I knew I had jumped my brains out, but would I appear to them any different? Especially now, the day before checking in with them, I wanted to behave.

Still, as my lunchmates made some final chitchat, it became clear to me that five of the eight miniature cookies—yes, I was counting—would not be consumed. As we stood up to leave, I asked them: Didn't they want to take them with? No, they each said. And as we walked together out of the restaurant, I looked over my shoulder wistfully.

•

I was indeed a slenderer version of myself. My weight had dropped to 182 from a little over 200. Polly did some flesh-pinching with her calipers; my body fat, she found, had correspondingly dropped from 20 percent to 11.5 percent. I now had less body fat than 90 percent of 30- to 39-year-olds, she told me.

"You look just great," she said, and indeed I felt as if I had stepped out of a Weight Watchers television ad.

I told her and Jamie about my workouts. A lot of squatting, I said. And a plyometric workout called the Air Alert system.

"Air Alert! Some of my buddies did that and swear by it," Jamie said.

"Really?" I was pleased to hear I wasn't completely wasting my time even as I was jumping myself silly.

"One of them gained something like ten inches."

"No way," I said. Wow. A man could dream. "So how old was your friend when he gained those ten inches?"

"Eighteen."

Phooey.

Jamie said he'd quickly put me through the same tests they checked through months earlier, to see if I had made any progress, carefully examining my stability, balance, and flexibility.

"Awesome," he kept announcing. "Much better." "Like Cirque du Soleil," he said at one point, and I could feel my cheeks redden. I was able to do at least one of those push-ups with the thumbs by the hairline, though Jamie had to gently prod me to move my thumbs higher up. "All the way to the hairline," he said, thus reminding me, unwittingly, that I continue to bald.

My overall test score in the Functional Movement Screen, the examination of my flexibility and power potential, jumped from 11 to 17. But my single-leg squat remained wobbly, Jamie noticed, and my hip and knee tended to fall out alignment. He pledged to put together some exercises that would solve these problems, or at least compensate for them.

Now it came time for the vertical. Was I actually improving? All those hours of exercise, of dieting, of misery: Did they add up to anything?

As I had months earlier, just before the start of my dunking proj-ect, I wrapped a bit of masking tape around the fingers of my left hand, stood next to a cinder-block wall, squatted down, and jumped as high as I could, spanking the tape against the wall at the highest possible point.

The result: I had jumped nearly 2 inches higher than last time. Just over a 10 percent increase.

I was thrilled, as if, blindfolded, I had just smashed a piñata. All that work, for so long, in virtual darkness. Two inches ain't much, but it showed that at least I—and, by extension, just about anyone past his or her prime   had the capability of improvement. Of course people grow fitter if they work at it, but to push yourself higher into the air than you ever have before suggests something majestic about our potential. "If I seem to boast more than is becoming," Thoreau wrote, "my excuse is that I brag for humanity rather than for myself." Yes, he was talking about our overlooked ability to head out and build one's own tight-shingled shelter—not to jump higher. But the point holds.

"How much time do you have?" Jamie asked me.

"I'm aiming to finish it in late August," I said.

"This August?"

"Well, yes, or the very end of summer," I said, as if that would buy me any more time.

I returned the following morning for a repeat of the dreaded Wingate test.

"Prepare to go balls-to-the-wall," Polly told me, as I mounted the stationary bike.

I corrected her: "You mean ball-to-the-wall."

"Right," she said with a smile.

We bantered for the first five minutes, as I warmed up my legs. Mostly we kvetched. She told me that she couldn't stand gyms, either, especially women who aimed for a "gym-fit body." "What's the use in it?" she said. "I mean, show me what you can do with your body." She was a strong, thin woman who walked leaning slightly forward, as if she were battling a stubborn wind. "I just want to be outside, even in a slashing rain." She said this all with the broadest Massachusetts accent—she grew up just outside Boston. She had been preparing for a bicycle race, and she told me, with a shake of the head, about the would-be ironmen who abandon their spouses and children to train fanatically. She was interested in the good that fitness could do, which is why she had recently partnered with researchers at Memorial Sloan-Kettering to examine how exercise improved chemotherapy recovery for women with cancer. She told me this partly because she knew I was a cancer survivor, and partly because I knew that she was a cancer survivor.

"Ten seconds," she then said, giving me a heads-up that my life was about to become temporarily miserable.

Five.

Four.

Three.

Two.

One.

The resistance ratcheted up, as if I suddenly had cinder blocks for shoes.

"Go! Go! Go!" she shouted as I surged forward—at least, as forward as one can on a stationary bicycle. I felt as if I were running up a down escalator, one moving increasingly quickly. I was trying frantically to pick up my legs, and the thirtieth second of my thirty seconds appeared ever further away.

"Done!" she cried.

I couldn't talk, only breathe heavily.

My peak power, she told me as I ceased gasping, was 885 watts—or more than 10 percent better than my previous peak power output. Judging by a series of reports on Wingate test results from around the world, I now most closely resembled the peak power output of a Saudi elite soccer player. Random, yes, but I liked the word "elite." Her eyeglasses now on—she usually hung them around her neck—Polly silently scrutinized a graph she'd printed out that mapped my performance during the test.

"What is it?" I asked.

There wasn't a spike and drop-off in my performance, just a steady decline over the final 20 or so seconds.

"That tells me you've got a slow-twitch muscle makeup," she said, with a drift of sympathy in her voice. I would need every fast-twitch fiber I could summon to dunk; slow twitch muscles wouldn't help me much.

I felt like a pruned tree, deeply rooted, no matter the shape or reach of its branches.

•

That night, I went to my 102-year-old grandmother's for Passover. She lives in Forest Hills, Queens, the land of grandmothers, in a semi-detached house on Clyde Street. The house had disintegrated like its occupant: the bricks long unpointed, a shifting foundation stiffening the door until it barely opened, and the front garden barren, given over to the sort of rocks one lays atop a gravestone.

When I had seen her last, in October, she still walked with a walker, and we had long conversations about the old days, about Vienna, about the old family farm in Nitra, in Slovakia, about the cruises along the Danube and the Black Sea that she would still like to take. She had long avoided any regular medication, and only a few years ago, after her last serious fall, a doctor told us she had the "blood of a twenty-nine-year-old." But her condition had slipped. As my aunt and I prepared the matzo balls and laid out the gefilte fish, my grandmother, now unable to use her legs, remained seated on a cushy chair, often appearing to be asleep, an oxygen tube snaking its way into her nose.

I roused her, to tell her about a lunch that week with my agent and editor. "They even paid the bill!" I told her.

"A big shot," she laughed. I asked her if she had done any jumping in gym class in Vienna. We're talking 1916 here. (She had always been good with specific memories, and her stories went beyond the borders of old photographs. When I grumbled about some math homework once, she confided to me that she had cheated on a school exam by sewing sine and cosine equations into her stockings.)

"No, I don't jump now. I'm no spring chicken."

"No, no, Mamama." I leaned in. "WHEN YOU WERE A CHILD—DID YOU JUMP AT ALL?"

"Sure, we did it. We did running and jumping. For recreation. But it wasn't all that popular. It wasn't like it was something the Rockefellers did, or someone famous."

She meant Rockefeller the robber baron, not Rockefeller the governor; it was the sort of throwback reference that dated her, as if it wasn't already obvious that she was really, really old. Did my project sound absurd to her ears? My grandmother had always been a largely unsentimental person. In her cupboards, glass Yahrtzeit candles, lit on the anniversary of her husband's death, won a second life as everyday drinking glasses. To spend a year jumping must have sounded quite frivolous, and maybe a little too American, to a frugal refugee. But if she had misgivings, she never let on.

We had the seder in the shadow of her old vitrine, with its glass doors and mirrored backing chockablock with European knick-knacks. It housed a crush of objects, the remnants of a Viennese life long lost—hand-painted teacups and heavy crystal bowls and art nouveau silver platters. They were pressing reminders, too, of all the people who didn't get out; for every Kiddush cup, so many relatives. Here and there, stuck in, were 1960s and '70s travel souvenirs, chintzy Mexican tchotchkes and patterned Japanese chopsticks, relics of failed efforts to move beyond the chasm of the war. As the meal wound down, my grandmother announced that she wanted to retire. Her helpers removed her to her room, into her La-Z-Boy, and we crowded around her, taking the dining chairs with us.

It was a crummy room. Plaster was half-falling from the ceiling.

The bank of windows was sealed with a plastic tarp and duct tape to keep the bitter winds at bay. Her bed, the same one her husband had died in of a heart attack 30 years earlier, was parked against a wall. Hanging above it, a faded photograph of my grandmother's parents and grandparents, one she was now too blind to see. A commode sat in the middle of the room.

She lay assembled in the La-Z-Boy small and fragile as an ancient doll, a ridiculous green fleece blanket, decorated with frogs, tugged about her. Her right eye was milky, a cloud permanently stationed in its blue sky, and her face was cracked and chipped. Her silver bob, lovingly washed by a nurse, was soft and almond-smelling. The EverFlo oxygen tank kept a regular rhythm as we began "Dayenu," a Passover song passed through generations. If He had delivered us from Egypt, it would have been enough. But he also gave us the Torah. If He had given us the Torah, it would have been enough, but He also gave us Moses. And so on. My grandmother, half-there, half-mumbled along. This is when she customarily roused herself, for so long the old lady impressing our Pesach guests with her relentlessness. Even now she stirred, driving through the lyrics with her parched lips. My father held her hand in his own, caressing it with his long thumb. My throat caught as I sang the familiar words. Whatever potential she might have wrung from life, whatever deep possibilities lay within her, were now exhausted. Perhaps her talent had been resilience. If that was the case, she had made full use of it, passing through a life of mothering, of war, of flight, of resettlement, of remaking herself as an American. Seeing her there, shrunk down, the comparison was obvious—her body was reaching its limits, just as I was pressing mine. The self-centeredness of the thought upset me.

"Mokele! Mokele!" she cried suddenly, as if she were beseeching the Messiah. It was the name she called my father when he was an infant. "Where's my Mokele?"

"I'm right here, Mom," said my father, and he reached over to gently stroke her forearm, the skin as thin as tissue wrap.

"Mokele," she whispered to him, "I want Mama." Still desperate, in her last days, for her parents, dispatched to Auschwitz seven decades ago.

That was the last time I saw my grandmother. Rebecca and I flew back from Texas a week later for the funeral. At the service, my father read a letter my grandmother had sent her aunt in March 1941 from Macon, Georgia, where she and my grandfather (and my father, a toddler) had been resettled after escaping the Third Reich. My grandmother's parents were trapped in Europe, and she was desperate to get them out. "It would have been better for me to die there, than always to exist with such trembling here," she wrote. By August of 1942 she had grown frantic: "I have constant and terrifying worries about my mother and there is no rest or pleasure for me," she wrote in another letter to her aunt, in New York. "I think of Mama always and the circumstances that bind her. I think of her when I eat, when I sleep, whatever I am doing. I cannot get any reliable information at all."

My grandmother was an indomitable woman, always looking for solutions to problems, and in my pew, hearing her helpless desperation, I buckled and wept. "I know my mother," my father wrote in his memoir. "She must have considered every possibility, everything that was remotely available to her in a world in which distance, borders, histories, time, mindless, hateful strategies, were conspiring against her lone efforts." It wouldn't be till years later, after the war, that she learned of her parents' fate, and those of at least 15 of her uncles, aunts, and cousins. She mourned for them the rest of her life, decades after they themselves might have had the blessing of dying a natural death. "One thing was certain," my father continued. "Because my grandparents did not escape from Vienna, my mother did not either."

•

As I stepped off the plane and onto the Austin jetway, a blanket of warm early-April air tossed itself about me. I felt like I had alighted on another planet, one far from the sorrows of New York City. With nearly five months left, I resolved to approach my project with a renewed earnestness: My grandmother, after all, would have liked me to make the most of what I had inherited. A few inches, by my reckoning, were all that separated me from the dunk.

11

Psyching Myself Up

99 days to go. *Double digits! 178 pounds. I start the day with a set of sit-ups at home, and then three sets of 30 push-ups. Rebecca, cross-legged in her pajamas, a bowl of cereal in hand, watches me from the couch, our scruffy, monkey-faced dog curled up beside her. "I'm impressed, baby," she says, sounding oddly genuine. "You used to do those all floppy, if you could even do them." I lie on my chest, panting, pulling together the energy to get up and go to the gym.*

A muggy evening in late July, 1996. Atlanta. As track-and-field events wrap up for the day, eyes at Centennial Olympic Stadium, including those of President Bill Clinton, fall on the narrow-bodied leapers. Charles Austin squares off against his rival, a rangy, pale, goateed Pole named Artur Partyka. At stake, high-jump gold. The height is now set at 7'9". Partyka clears, even as his shorts lazily graze the bar. Twice Charles has failed, his heels clipping the obstacle. He was trying too hard to blow the entire jump up, to jump as high as eight feet. Charles turns his mind to his kids in Hawaii, and to what he is going to eat after the competition. It's his final chance to clear 7'9", and, in a move that speaks of his bold-ness and his calm, he orders the bar to be set at 7'10". He smiles and shakes out his shoulders. He approaches the bar with religious con-viction. Confident, aggressive—and yet under control—he bounds forward in the familiar J-shaped pattern. He plants his left foot and springs, quickly twisting his body in midair as he sails over the bar. He lets out a roar, puffing out his chest, thrusting his fists down to his knees as he stands atop the slouchy high-jump mat—the mo-ment of Olympic gold. Now it is Partyka's turn, now his final run-up, his final chance to match Charles. He looks as if he's about to cry, like a thin, sad, very skinny, very tall Jesus. He fails.

●

Charles told me this story in his gym, after I'd failed for the third time in a row to dunk a mere tennis ball. If I couldn't manage that, how was I ever going to dunk a basketball? Charles looked at me and said: "You're too much like this"—and he tensed up his entire face and looked serious, just about the only time I didn't see him smiling. I knew what he meant. I could feel my brows knitted as I readied myself for yet another run to the basket. "You run faster and you jump higher when you're in the flow," Charles continued. "I had already jumped that height in my mind. Making it was nothing. It was play.

"Stop thinking," Charles commanded. "Just get up there and put that ball in the hoop."

And then, not long after I had returned from New York, and closing in on four months to go, I managed to dunk a tennis ball on a full-throated rim.

After I threw it down, Charles lit up.

"Hey, Asher, you have any kids?" he bellowed, even though I was only a few yards from him.

"No," I said.

"You going to go home and tell your wife you dunked a tennis ball and you're going to have a baby in nine months."

He let out a belly laugh.

"How's that feel? You can't stop showing your teeth, you're smiling so hard."

It's true: I was pleased with myself. Not yet a basketball—far from it—but I was showing some tangible progress. And I had only had a few sessions with Charles at this point. The man was getting me to believe in myself.

He was also making me stronger in just the right ways. Twice a week I was driving the 45 minutes each way to his gym—down through the suburbs south of Austin and finally off the highway just across from the outlet malls. Outside the gym, which looked like an oversized Quonset hut, stood Charles's black Camaro, surprisingly lacking a vanity plate, and the heavy-duty Chevy pickup of his assistant, a beefy guy named Terrell.

I usually showed up at 9 a.m.—I was on book leave from the newspaper now—and the gym was largely empty, save for me and Charles and Terrell and maybe one or two other people, a high schooler getting ready for the coming football season, or a retiree with time on her hands. Charles might have me strap on a vest with a pair of nylon ropes dangling from behind. These he'd grasp, standing about 10 feet behind me. "Pull me," he'd say, and like a yoked giraffe I'd start sprinting down a track, huffing and puffing, the weight of Charles's muscled body making it that much harder. "Get off your toes!" he'd shout, like a mantra. Lord knows I tried, as I felt my hamstrings on fire. I worried that some of that sweat flying off my frantic arms and legs would splash Charles. But he seemed unfazed. "Get off them toes, Asher!" He wanted me to drive off my entire foot, pushing my knee up each time as high as it would go, which wasn't much higher than my belly button. We made a funny couple, tethered together by nylon: I, pale, balding, hairy, effortful, inefficient, grunting; Charles, smooth-headed, toned, economical, powerful, easygoing, a chatterer.

Charles was an entrepreneur, often holding forth about the long-term plans he envisioned for the gym—how he hoped, for instance, to pass it down to his kids. He was only 46, but he had come from a large household of scarce opportunity and didn't want to see the success and reputation he had won just fade away. (The only time he'd leave the gym was for a business appointment or to have lunch with his wife. "What did you eat?" I'd ask, at an early-afternoon training session. "Nothing. I just stared into the eyes of the most beautiful woman in the world," he'd say.) His latest idea was a simple, ingenious piece of exercise machinery that he hoped would make its way into the home of every middle-class American, like a NordicTrack of its day. (My parents had a NordicTrack. I'm honestly not sure they used it even once; for a half-dozen years it served as a pricey towel rack before they finally got rid of it.) In a nutshell, the invention, a rectangular piece of fiberglass, allows for a version of stationary speed-skating. Exercising on the machine was hard, as a matter of coordination, and, of course, tired out my hips. I asked him if he would be doing infomercials: "Everything, man, everything," he said, and as he got on board the device to show me how it ought to work, smoothly, effortlessly gliding back and forth, I thought: "Even though I am fully aware this would end up an expensive towel rack in my household, I'm prepared to buy this exercise device from this incredibly charismatic, confident, and handsome man."

I may have had a little crush.

A stopwatch hanging about his neck, Charles ordered me to undertake slalom shuffles, squat jumps, resistance-banded walks. These were intended, as Jamie had once explained, to recruit my glute muscles. "Recruit" was one way of putting it; "make miserable" would be another.

And, worst—or best—of all, he ordered me to jump on his four-foot-thick high-jump mat. Have you ever tried standing atop a high-jump mat? I hadn't, either: It's like a foamy water bed. Even getting on top of it is hard. It's like an adult version of one of those ball castles that delight children. Charles would slouch atop a plyometric box, stopwatch now in hand, and order me to jump in place, two-footed, for a minute. Sweat would go everywhere. Legs would burn like roasted chicken. I'd break for a minute, walking about the big mat like a drunken sailor; jump some more; break for a minute; jump some more. Then he'd have me jump backward, or hop, or side-shuffle on the mat. By the end, my shirt would be fully drenched. I'd stagger off the mat and stab at a towel I had brought along, drying myself for the fifteenth time that hour.

Charles knew I was a journalist, knew about my project, and had jealously guarded his regimen. He worried that I would reveal too much.

"I'm invited to conferences and they say they'll pay me thousands of dollars," he jabbered in a monologue as I slogged away on an exercise bike—fast rhythm, heavy resistance, short ride—too exhausted to do much more than nod. "But I don't want them to have my secrets.

"Terrell, how long you been working here?" he called over his shoulder to his second-in-command, a jolly 6'5" dude who grinned and lumbered his way around the gym.

"Eight years."

He turned back toward me. "I tell him only what he needs to know."

The weird part was that nothing Charles was instructing me to do was novel: box jumps, high knees, clapping push-ups, hurdle jumping. Pushing myself to please him made the time so valuable.

"I'm like a doctor healing a bad patient," he had told me, in one of his frequent koan-like pronouncements. (But really, did he have to liken me to a sick person?) "I want to see you make the most of your potential. It makes me feel good." And I, in turn, wanted to be a model patient. It says something about me—an unhealthy deference to authority? a precocious obsessiveness?—that as a boy, I had always aimed to be the perfect customer at the Belnord Unisex Barber Shop; my goal, neurotic as it sounds now, was to move my head this way and that before the barber ever had to give me direction. I never quite anticipated every command, of course, and I remember my small deflation when the barber, standing behind me, inevitably ceased the *clip-clip-clip* of his scissors, and said: "Straighten up, please."

A couple of weeks after the tennis ball throw-down, my initial warm-up sprints completed, Charles handed me a squeezy yellow ball, the sort we used in dodge ball as kids. It was a little bigger than a softball, a little smaller than a volleyball. About three times the size of a tennis ball.

"Dunk this," he ordered.

I wasn't alone in the gym. Charles' eight-year-old son, Cameron, was shooting at a rim on the other end of the gym, and his 19-year-old son, Allex, the state high school high-jump champion, was working out on an exercise bike.

I tried a half-dozen times to jam the yellow ball, but I couldn't do it: The ball hit the front end of the rim or I threw it off the back iron. And now two generations of the Austin clan were staring at me. Partly, I just didn't want to disappoint Charles. But partly I didn't want to disappoint myself. I was angry at the rim. Or maybe, if I'm honest, I wanted to be angry. I can do this, I told myself. I thought about those weightlifters at the gym who scream during their final repetition. I'm not even screaming, I thought. Perhaps I'd feel self-conscious, a poseur, summoning a show of passion for a purely physical act. My brother-in-law Ben had played professional basketball until, he had told me, he decided he'd rather write than train. Of his playing days, he once wrote, "I never learned the trick of falling into a passion like you fall into step."

Charles could tell I was frustrated: "Close your eyes," he commanded, as I prepared to start my run-up. "See yourself dunking."

My eyes opened.

"Now do it," he said, as if he had performed a Jedi mind trick.

On my eighth try, to my great relief, I dunked the ball. With nearly a third of my dunk year left, I was definitely moving in the right direction.

Charles said: "You going to get home to your wife as fast as you can."

Allex hunched over his exercise bicycle and mimed swerving a steering wheel with his hands: "He's going to be driving like he's got a race car!"

"What'd she say the last time you told her you dunked?" Charles asked.

"She smiled a big smile," I said.

"Yeah? How big?"

"There was lots of smiling going on," I said.

What Rebecca had actually said was: "Maybe I don't want your non-dunking genes polluting my dunkers' gene pool."

•

"The limits to human performance that we now perceive do not represent physical realities so much as they signify failure of the imagination," John Jerome argued in *The Sweet Spot in Time.* "Those limits don't really exist; they are ghost images, lying there waiting for us to surpass and dissolve them."

These "ghost images," I was learning, are what top athletes and coaches try to banish. "I don't like the word 'limits,'" Todd Wright, the burly University of Texas basketball strength coach, told me when I first interviewed him. He had called me back only reluctantly, he said, because he was uncomfortable participating in a book that even considered limits. "The first thing we teach our players is the only thing they should doubt is their doubts."

I'm sure great athletes, and performers of all stripes, have greater confidence than the rest of us, at least when it comes to their field of expertise, but I could perceive a cult of limitlessness appearing before me. "A couple of words we never use are 'can't' or 'won't,'" Will Lenzner, a trained sports psychologist who has served as the director of mental conditioning at the Chris Evert Tennis Academy, tells me. He has trained martial artists, golfers, and military snipers. ("When they come to me they've either hit rock bottom or they're just below the top and want that extra edge.") "I don't believe in the word 'try.' That gives them an out. I don't care if you dunk a tennis ball. I want the attempt to be successful. I don't want you thinking about dunking it. Think about the process: three steps, one power dribble, and boom! We focus on dunking, and dunking becomes an insurmountable goal. But we focus on those three power steps and we're much more likely to find ourselves dunking."

•

For a time, Lilo, a petite, white-haired, 50-year-old Austin attorney, kept a small pile of her broken wood at her office, a neat form of intimidation, until she needed to reclaim space on her desk. Michael, a brown-belted, tubby 14-year-old with a premature moustache, keeps his neatly in his closet.

The cracked timbers were the leftover evidence of cocksure flying fists and heels, souvenirs of training at Kim Geary's karate studio.

"I want y'all relaxed but serious, confident but not cocky. Everyone here is more than ready," Master Kim, as she is known, told the half-dozen barefoot students assembled before her. They wore white robes, each cinched with a colored belt; she, black on black, with hair short and curly like steel wool. She had a rugged face, with deep lines setting off her cheeks. "If you feel nervous, just take a deep breath."

The task before them: breaking through stacks of white pine, with a combination of kicks and chops: spear-hand; rock-knuckle fist; ridge-hand; palm-heel; knife-hand. I had wondered about the mental effort it takes to do the seemingly impossible, which is how I ended up at Master Kim's studio, all blond pine and mirrors and windows, sheltered on a small street from a cold, rainy Texas morning. Who in her right mind would strike a piece of wood? Or a fired brick? It seemed to me a pure execution of mind over matter. If these people could convince themselves to break objects with their bare hands, if they could conjure up the necessary mental willpower, shouldn't I be able to persuade myself to dunk a basketball?

Before the collision of bone and timber, Master Kim, 53 years old, an eighth-degree black belt, led her students in a series of chants: "Respect others. Refrain from violent behavior." I thought for a moment that maybe this would be the milquetoast of martial arts, that I would see little of the mental and physical stretching of potential that I was hoping for. Then came the martial-arts-style calisthenics. "Four-knuckle fist to the throat," Master Kim barked, and the class of six, in synch, would punch through the air with a shout, as if engaged in a *West Side Story* street fight. "Kick the belt, punch the face," she ordered. "Kick the belt, chop the neck." I imagined an unhappy adversary, wobbly now, bloodied, who picked the wrong bathrobed sextet to mess with. "Two punches to the face, roundhouse kick to the temple." They sparred with martial arts styles called "Destroy the Fortress" and "Escape the Mess," which made me think of those old Bruce Lee movies where he had to overcome dozens of no-name bad guys on the chief villain's unmarked island retreat.

Then, with everyone rapt, Geary laid a red brick across two cinder blocks. "A black belt demonstration," she announced. She knelt on one knee to the side of the brick and pulled her right hand high above her, slowly, steadily bringing it toward the brick a couple of times, as if measuring out a golf putt. For most of us, breaking a piece of pine with our bare hand would seem a fool's errand; breaking a brick seems positively counterproductive. And when I had introduced myself two weeks earlier and explained why I had contacted her, Master Kim told me that she would not be cajoled into breaking anything. She had not broken a brick in a half-dozen years—when you're a badass eighth-degree black belt, what do you really have left to prove? But I wasn't so surprised that rainy day when she knelt by the brick and prepared to strike it. Geary's career had been built on proving herself: She had begun her karate studies in earnest in Houston in the mid-1970s, under a sadistic master who forbade the drinking of water during workouts and encouraged his students to run barefoot on a scorching-hot track nearby. All the other students were male. "I had to be better than the guys," she explained.

Now she pulled her hand back once more. And then— Haaiiieee!—she swiftly brought it through and cut the brick in half.

"It's been a long time," she said, to no one in particular, standing up and smiling with some relief as her students hustled to remove the cinder blocks. Later, after the class ended and the students had all changed and gone home, she confided in me. "I dreaded it all week," she said.

●

What is it about Master Kim or Charles Austin that allows them to dream up a task and then make their bodies follow? As with a kid spotting up for yet one more shot on his driveway rim, their motions become more a knack than an effort. They grow fluent in the language of their limbs.

Most of us lack that kind of intimacy with our physical selves. It's a luxury of middle age: Old people and athletes are much more attuned to their bodies than the rest of us. Our bodies seem only loosely attached to us until something—a tumor, perhaps, or a clogged artery—reminds us with a jolt that we're bound up with them. Josh, my eldest brother, a marathoner and vegetarian, thinner and taller than I, woke up a few months ago to an unexpected, alarming shortness of breath. After a trip to the emergency room and a battery of tests, the doctors diagnosed a congenital calcification of his arteries. "I feel like my body betrayed me," he told me. (I got the news on the road, during a newspaper assignment, and it shook me. He's my dependable big brother and, in a way, I felt like his body had betrayed me, too. He asked me to tell our parents; as I dutifully did, sitting on an old, high curb in the empty Texas town of Gonzales, I suddenly was launched back to my revelations about my own cancer: I found myself desperately trying not to fall apart.) Our bodies are foreign to us—hair, teeth, fingernails, both alive and dead at once, involuntarily growing out of us. (It's a feeling that stalks us during puberty, as if we were strangers to our own bodies, meeting them for the first time. Some people are even convinced they were born the wrong gender.) In the throes of the flu, or the fogginess of a bad cold, we wish we could trade our bodies in for an upgrade. This flesh and blood happens to house our brains, and the most we can hope for is that the foundation holds up, that the plumbing doesn't clog, that the halls remain tidy, that the windowpanes don't break. We see ourselves in photos and wonder how true the likeness is—because it doesn't look the way we imagine ourselves, just as our voice, played back on a message machine, doesn't match the way it sounds in our own head. Part of the reason somebody might be described as "comfortable in his own skin" is precisely because so few of us are actually comfortable in our own skin.

Yet our alienation is a mild one. If on one end of the spectrum athletes, karate masters, and ballerinas, mind and body in harmony, coax their bodies to clear high bars, break bricks, or dance *en pointe*, at the other end of the spectrum a motley crew negotiates chasms far greater than our own, between their psychological sense of themselves and their actual physical ability.

In 1874, a young French psychiatrist was summoned to consult about a 43-year-old woman admitted to an asylum just southwest of Paris. The problem: the woman was convinced she essentially did not exist. Mademoiselle X told the doctor, Jules Cotard, that she "did not have a brain, nerves, chest, stomach or guts," he recounted in an 1880 lecture, in which he identified the illness as *délire des négations*—an attitude in which the patient denies the existence of the self. Of course, Mademoiselle X did have a brain, a chest, and a stomach, even as she insisted "all she had left was the skin and bones of her disorganized body." Though rare, cases of the delusion, now known as Cotard's syndrome, or more colorfully as walking corpse syndrome, still crop up. In 2005, Greek doctors reported on a case in which a 46-year-old woman, twice jilted and swindled by fiancés, was escorted to a hospital by two of her brothers. She informed doctors that "all the organs within me have rotted. . . . I am tired, I haven't slept for years, I have no blood, I have no heart, it doesn't beat anymore," she continued, in words as poetic as they are troubling. "You are deceiving me when you take my blood pressure, because I'm not alive anymore, I'm a dead plant."

You can think of Cotard's as the reverse of phantom limb syndrome, in which amputees report feeling tingling or other sensations where a limb no longer exists. Doctors believe Cotard's, a cousin of other paranoid delusions involving identity, results from a combination of deep melancholia (which is partly why it's so rare nowadays, since depression can often be treated) and, oddly enough, profound, hard-to-treat facial recognition problems. A glance in a mirror reveals a total stranger.

Often, as with Josh, who is now fine, the source of that anxiety, a betrayal of the heart, is all too real. But the profound mental disassociation of the Cotard's patients reveals how important it is that we convince ourselves, consciously or unconsciously, of the capabilities of our bodies. Few of us overestimate our physical potential; most of us underestimate it. For world-class athletes, on the other hand, there is little daylight between their sense of their athletic capabilities and their actual capabilities. To put it geometrically, for them the two are congruent. For the rest of us, bridging that gap has great consequences for how we perform, whether the aim is to sing or to dunk.

June 10: 76 days left. *A shaved-head, box-eared guy stopped me in the Greek yogurt section of the supermarket today. "You work out at gym?" he asked me. "Yeah, I do," stunned that whatever I did at the gym had caught anyone's notice. "I'm Vasily, Russian boxer. Live here now with wife. I watch you work out. You work out hard. You make legs strong." Russian-boxer-dude just complimented me on my workouts. I'm squatting 190 pounds, plus squat jumps with 65 pounds across the back. (I know, I know—most of you can do that in your sleep. But for me, inflexible as I am, this is a big deal.) But my right knee is hurting more and more, for longer and longer, after each workout. I ice it down regularly. Weight: 175 pounds.*

Progress report: As my jumping was coming along, I wondered about the two kids I had met months earlier, Laquan and Josh, who appeared to be on their way to dunking.

Last I had seen Josh Scoggins, at a Hill Country gym, he could dunk a volleyball. When I wrote him a few months later to ask after his progress, he told me his dunk project had stalled because he had been suffering severe pain in his knees. Doctors had diagnosed him with Osgood-Schlatter disease, a sort of temporary arthritis not uncommon among adolescents growing into themselves. He had to lay off the jumping, he told me.

Laquan was not faring much better. His mother, Demetria, had invited me to come out and watch him play in an after-school league. So on a Tuesday night I headed back across acres of farmland to Pflugerville, the suburb with the funny spelling that had become home to a lot of middle-class African-Americans, the consequence of black flight as Austin neighborhoods that had been African-American since the Civil War grew brown.

Laquan was the biggest player on his team, but he started on the bench, in one of the plastic folding chairs by the coach, a comb-over in a white polo shirt and glasses who was earnestly spewing instructions to his team at a rate I'm pretty sure most of them couldn't process. Laquan looked unhappy. I also noticed he had a bandage wrapped on his left hand. He was still wearing those beat-up kicks; the other kids had gleaming white sneakers.

I was standing with parents on the other side of the court—Demetria was off looking after her grandson. One couple was actively heckling players on Laquan's team as they tried shooting free throws. It seemed unfriendly to me, a weirdly hostile thing to do in a meaningless after-school league with middle schoolers. A guy beside me, presumably a parent, was paying no attention at all, squatting over a copy of *Front Sight: Your Leader in Firearms Training.*

The comb-over didn't put Laquan into the game till just after halftime. By now I had found Demetria, who told me that Laquan was in no mood to talk. He had been in a car accident the day before and had been shaken. He had also sliced his hand, which explained the bandage. Now he was sullen. Maybe that explained why he was on the bench the first half. Things hadn't been working out so well for him, she said. The Pflugerville school district had been paying for Laquan to visit Chris Corbett's basketball camp, where I had originally met him, as a sort of scholarship to give him some structure and mentorship. But the district had cut the scholarship money since I had last seen them, she said, ending the trips to the basketball camp.

When Laquan got in the game, his team was trailing 31 to 29. Maybe now, I thought, given an opportunity to play, he'd go to work. But he seemed disengaged, trailing his teammates when running back for defense, unwilling to grind into a good position down low on offense. He was loafing, really, a word that I remembered Estes Banks, my elementary school soccer coach and a former Oakland Raiders running back, using to describe players who didn't fully commit themselves to the punishing drills he had prepared (among them, running in place, in the mud, until Estes screamed "Hit it!" and all of us, many of us lawyers' and doctors' sons, dived to the ground and picked ourselves up and ran in place more, as if we were in Raiders training camp): "Loafers," he called us. (I was known by my teammates as "The Professor" because of my tortoiseshell glasses.)

"We need you, Laquan!" a teammate's father called from the sideline.

Laquan got fouled on a move to the hoop. Then, suddenly, he thrust his chest out at the defender and then shoved him with two hands. He was losing control. And the refs whistled a technical foul.

Front Sight guy snapped his head up, for a second, from his glossy firearms mag.

"Get him pissed off!" screamed the heckling woman, in a jeer at Laquan.

"Clear your head, Laquan!" called the father of the teammate.

The coach pulled Laquan aside and, I could tell from his body language across the court, was trying to calm him down. He kept pushing his two hands downward through the air. Laquan chilled out, but then, again, he seemed disengaged, beaten downcourt by shorter players. When the game ended, Laquan's team had lost 43 to 31.

As he walked to his mother, I asked if he was up for talking some. He just shook his head. And he looked like he might cry.

I had assumed these kids might jump farther and higher than me easily enough, but their potential, it seemed, had its own limits. Someone once wrote that a young man is a theory, an old man is a fact. These kids were learning that we were not all quite created equal—something dunking distilled quite neatly. Figuring out their capabilities was a puzzle Josh and Laquan had not quite solved, a test of confidence, of maturity, of physical durability.

●

Little more than a half-dozen years earlier, when I was subjected to another round of cancer toxins, a nurse told me that mine was a chemo cocktail concocted to be so potent that only young men could withstand it. Even so, I had felt bad for the patients around me, women chiefly in their 50s and 60s, whose cancer treatment was open-ended and uncertain. The most many of them could hope for was a long-term temporary reprieve. I knew I had a different trajectory. Two weeks after my initial diagnosis, I prepared to travel to Indianapolis, to begin my treatment with Dr. Larry Einhorn, the researcher who had turned the tables on testicular cancer. (Still, some people, including the son of a family friend, another Einhorn patient, die of the disease.) He had told me to bring everything to Indiana, including a sheaf of CAT scan films and my forensic report. "Don't check anything in," he told me. "Carry it all on." Rebecca and I drove by the Austin hospital to pick up the report: inside a small bubble-wrap-lined manila envelope labeled "fragile" was a test tube with what looked like a half-consumed sucking candy—a cross-section of my testicle. My left nut, so recently hanging inside my scrotum, now flew snugly in the seat-back pocket ahead of me to Indiana.

I still remember Einhorn, who had the demeanor of the nice Jewish doctor he is, standing by a window and holding up the films to the thin Indiana light. He turned to us. "Your cancer is one-hundred-percent curable," he said. And once he rid my body of it, he continued, "the odds that it will return are about the same as my crossing the street outside and getting hit by a car." Rebecca and my mother, who had flown in from New York City, embraced in relief; it was the first good news we had gotten in weeks, and it had come down from this Moses-like figure. But the treatment, he warned, was a brutal one. "If you're ready, let's get started," he said. "Of course," I said. "Let's do it." We went up a floor, from the medical offices to the cancer ward, and the task ahead of me seemed much more real. I stripped down, put on a medical smock, and settled into a grayish-green dental-style chair, surrounded by IV stands and beeping equipment. A nurse prepared a needle, and as she stuck it in my left forearm, I promptly fainted.

12

Aiming Too High

57 days left: *107 degrees today in Austin. A wet, nasty late-June heat. Two sets of 10 by 100 sprints, with 45 seconds and six push-ups between each sprint. And five minutes between the two sets. High arcs of sweat went flying from my fingertips each time I pumped my hands. My shirt was soaked through by the sixth sprint. Weight is still at 175 pounds. My chest has grown a couple of inches, my posture has straightened up, my calves and thighs are thicker. But I can't wait for this year to end—my knee aches any time I drive for more than 20 minutes.*

His butt glued to an office chair, Edward Coyle wheeled himself toward me, hitched up the left trouser leg of his khakis, pulled down his black dress sock, and thrust out his shapely calf.

"See those little white marks?" he asked, and I found myself bending over, nose inches from his handsome gam. Between the lines of black hair I could indeed spot dozens of thin scars. "Those are from the biopsies, thirty years of them."

We were sitting in his lab, the same one he had occupied for 32 years, a low-ceilinged, cinder-blocked, windowless space, buzzy with fluorescent light, on the eighth floor of the grandly named Bellmont Hall, really a concrete block beneath the bleachers of the University of Texas football stadium. ("We call this 'Bellmont Bunker,'" he had told me by way of sardonic greeting.) Feet away sat a row of stationary bicycles hooked up to a bank of desktop computers. Plastic tubing, like giant bendy straws, was strung from some of these computers to the ceiling and then dropped down in front of the bikes.

In his pressed khakis, button-down short-sleeves, clunky black sneakers, and immaculately clean spectacles, he resembled a genial police detective more than a former running wunderkind. The only trace of a faraway past was the faint Queens accent, an accent I could place only because it sounded an awful lot like the voices of the super, the grocery-bagger, and the nurses who had helped my grandmother for so many years. He had been a stubby Catholic kid from Woodside, attending Mater Christi, a school that catered to both boys and girls—in separate classrooms, of course. The cafeteria had a divider down the middle, and you could hear girls' giggles on the other side. He had no jones for running, but in a forced P.E. run around Astoria Park in his freshman year, he scorched nearly the entire class of 250. His homeroom teacher was also the track coach and, in due time, coerced young Edward into quitting the swimming team, not that, being small, he had ever been particularly suited for the pool. Eventually, as a student at Queens College, he led the track squad in everything from the 800-meter run, a searingly long near-sprint, to the 5K, an event pregnant with psychological pressure and strategy, winning New York City titles in a clutch of distances.

It was in college that he began thinking about what separated his performance from others. How much of his track ability, he wondered, was due to natural muscle composition, and how much of

it was molded by relentless training? His workouts became a voca-
tional pursuit, and Coyle became his own guinea pig. The muscle
biopsies would reveal, precisely, what portion of his fibers were fast-
twitch and what portion slow-twitch. Slow-twitch muscles use oxygen more efficiently to generate energy; fast-twitch muscles are less
efficient but fire more rapidly and generate more force. There is a
surprising amount of variation in the way these muscles are distrib-
uted among individuals. The leg muscles of someone like Carl Lewis
might be composed of as much as 80 percent fast-twitch fibers,
while the leg muscles of top-notch marathoners might contain up to
80 percent slow-twitch fibers. Meanwhile, the legs of average Joes,
of you and me, tend to fall into a 50-50 mix of fast-twitch and slow-
twitch muscles. A fast-twitch muscle reaches its peak tension—or
the point at which it's doing the most work—in a tenth of a second;
a slow-twitch muscle takes two-tenths of a second. On the face of
it, a tenth-of-a-second difference doesn't sound like much. But, of
course, one is twice as fast as the other, and as two sprinters bound
down the track, relaxing and contracting their muscles as quickly as
possible, each tenth of a second makes a big difference. What Coyle
found was that over time, as he favored longer events over shorter
ones, the slow-twitch muscles were building up—even as the fast-
twitch ones languished. Post-college, he was the anti-dunker, trading in the sprint for the lope.

But Coyle told me he was unwilling to put performance down
purely to muscle makeup. "Remember," he said, leaning forward in
his swivel chair, "we talk about *neuro*-musculature. It's not just the
muscle makeup but how quickly your cerebral cortex can send sig-
nals down through the spine to work." Training ourselves to send
quick impulses to our muscles to do work, especially the kind of
rapid work involved with fast-twitch fibers, is possible, he said. For
a runner like Coyle—a runner who only really got going late in college, with nearly seven years of instruction behind him—it wasn't

only muscle makeup that mattered. It was also the economy of technique, the ability to convert oxygen to energy. And, of course, the coaching.

Coyle told me a story about a group of heart-disease patients he worked with in the late 1970s at Washington University Medical School in St. Louis. These were 50-something men, recovering from serious heart problems—"Parts of their hearts were dead," he said. "They thought their lives were over." Coyle's job was to train them as intensively as they could withstand. Calling on his old running days, he began prepping them for long-distance jogs, starting with just a few hundred meters. Some of them were able eventually to run 10K races, farther than Coyle himself had covered competitively, and a couple went on to complete marathons. "It's amazing how much the body can improve," he said, shaking his head even now, beneath the gently humming lights of Bellmont Hall.

•

The notion that humans can improve in measurable ways is now enshrined in the academy, tested daily in labs like Coyle's: Just about every university today has some sort of department dedicated to kinesiology, a field built on the inherently optimistic, essentially American premise that humans have all sorts of room to improve on their apparent potential. An equality of physical opportunity. The field has its own journals and it gives out doctorates. And, of course, every high school in America seems to have its own weight room and its own phys-ed teacher, dedicated to making kids think they can transform their flesh, like their minds, into something useful. Or at least to making them sweat awkwardly for 25 minutes before showering off in a grotty locker room. Nineteen twenty-four, the year Dudley Allen Sargent, the inventor of the vertical jump test, died, marked the first time a course in teaching physical education was offered at a teacher-training department. By 1930, a quarter of schools in a national survey said they had a physical education program; none had one at the beginning of the century. In a sense, the buildup of the business and study of self-improvement is the benign legacy of an age of hubris, one that saw humans increasingly confident in their abilities to classify everyone around them. This was a period when education "was consumed with a passion for precise measurement," Harold Rugg, the now-forgotten progressive educator, wrote in his 1941 memoir *That Men May Understand.* "We lived in one long orgy of tabulation. Mountains of facts were piled up, condensed, summarized and integrated by the new quantitative technique. The air was full of normal curves, standard deviations, coefficients of correlation, regressive equations." At the outset of this age stood an idiosyncratic institution, the Harvard Fatigue Laboratory, whose elite researchers had the gall to carry out a singular mission: to pin down the breaking point of men. (They weren't particularly interested in women.)

•

In 1927 a plump, red-bearded, suspenders-wearing, pipe-smoking Harvard biochemist named Lawrence Henderson had an idea: He would collaborate with American corporations to research ways to wring efficiency from American laborers. Known lovingly by his colleagues as Pink Whiskers, Henderson was a pompous gourmand who favored the word "superior" and who compulsively ranked things as relatively meaningless as types of denim overalls. He saw himself as "a modern-day Socrates," one of his former colleagues later wrote; it was a role he performed beautifully. With young scientist-devotees clinging to him, Henderson led the laboratory on a series of tests meant to get at the empirical limits of human capability, as if a human were a type of metal with specific boiling, melting, and freezing points.

Henderson's right-hand man was, in many respects, his opposite: a workaday, crew-cut-wearing research scientist named Bruce Dill, who, orphaned at age eight, cut a character austere, subdued, Presbyterian. For his meticulousness, his objectivity, his honesty, and his curiosity, he was remembered by a colleague as a "scientist's scientist." He was also, like Edward Coyle after him, something of a willing human guinea pig, a kind of comic-book hero who urged other researchers to bypass the usual tests on chimps or other creatures and run their experiments on him. (In my experience as a one-time reader of comic books, that person usually turns out to be the headstrong villain—Dr. Octopus, for one, or the Green Goblin—but by all accounts Dill was a wholesome, patriotic American.)

Thus, after Dill and other researchers conceived of the 40-40-40 Club, to honor investigators who lingered in a chamber that simulated conditions of negative-40 degrees Fahrenheit and altitudes of 40,000 feet, and undertook arduous 40-mile walks, Dill himself, physically courageous and unflagging, became the first member. (He stuck it out in the "cold room," as it was known, for 20 minutes.)

Besides the cold room and a hot room, which could reach 115 degrees, the 800-square-foot lab, tucked into the basement of a red-brick colonial building on Harvard's campus, was equipped with an X-ray machine, standard treadmills, and, for God knows what purpose, a dog treadmill. Reading the lab members' research accounts, you can't miss their cheery, Sargent-like optimism in their own capacity to pinpoint the extent of human capabilities. What were our parameters, as a species, they wondered, and what happened if we overstepped those boundaries? The lab's researchers accompanied pilots on flights from California to Hong Kong to monitor their alertness, and collected the perspiration of sharecroppers in the Mississippi Delta. They studied the relationship between fatigue, recovery, and lactic acid accumulation among Harvard football players. They devised a formula—"the Hobbling Effect"—to calculate, in units of energy, the encumbrance of heavy garments worn in the snow. They established, after a spate of deaths during the building of the Hoover Dam, that sweat pouring off laborers in the Nevada desert had half the salt content as sweat produced by Bostonians during a Massachusetts winter, leading the researchers to recommend that copious amounts of salt be added to the workers' victuals.

By 1942, Henderson had died, and Dill had enlisted in the military. Competition for money within the university, territorial battles between the business school and the lab over corporate sponsors, and academic politics crippled the lab's work. Five years later, the Harvard Fatigue Laboratory was disbanded, its equipment and records absorbed by the school of public health. The lab had lasted only 20 years, but its alumni fanned out across the country, sowing the seeds of the lab's optimism to the Arctic, where a Fatigue Laboratory biologist studied circadian rhythms; to Florida, where a Fatigue Laboratory cardiologist studied the effect of weightlessness and acceleration on human balance as he prepared astronauts for manned space flights (his research led to the development of drugs to combat motion sickness); to the Army, where the lab's chief technician served on a committee called Nutrition for National Defense; and to office parks everywhere, whose chairs and desks were inspired by the research of a Fatigue Laboratory alum who is remembered as "the founding father of ergonomics." Encouraged by the example of the Fatigue Laboratory, universities around the country began establishing their own human performance labs.

•

Edward Coyle told me to clamber aboard one of the "power/ cycles"—stationary bikes hooked up to computers—that were standing around his lab. He had made a name for himself creating an abbreviated version of the Wingate Test to test the fast-twitch abilities of humans. He has counted among his clients the San Antonio Spurs basketball team, the Chicago Blackhawks hockey team, NASA astronauts, and, of course, University of Texas sports teams. With players tracked over the course of a season, he can look at the results to rank talent and guard against overtraining. "If that rating goes down, something's the matter," Coyle told me, as I slipped my feet into the cage-style pedals.

Nearly two decades earlier, Coyle's most famous client strapped himself into the very same stationary bike, sucked on one of those plastic straws, and blew the VO_2 max ratings off the charts. At the time, Lance Armstrong was just a brash second-tier cyclist from Texas, still unknown to the general public and still cancer-free. Over seven years, Coyle tracked Armstrong's improvement, growing as close as one could hope to an athlete whose eyes were on a prize much bigger than anything that could be had in Texas. In a 2005 peer-reviewed article in the *Journal of Applied Physiology* titled "Improved Muscular Efficiency Displayed as Tour de France Champion Matures," Coyle argued that Armstrong dramatically increased his power output by remaking his body after cancer and becoming a more efficient cyclist. Around the same time he wrote that article, Coyle served as a witness for Armstrong in a suit against Dallas-based SCA Promotions, which refused to pay him bonuses, citing doping suspicions.

"You tested Lance, right?" I asked Coyle.

"He was off-the-charts amazing," he said, as if the cloud that had cast a shadow on Lance's Tour triumphs had spared his research lab.

I could understand the unwavering loyalty to Lance. He had boosted my own spirits and offered some practical help in my darkest moment. The weekend I was diagnosed, in 2006, I telephoned my editor at the newspaper: I wouldn't make it in that week—cancer and all. Not a bad excuse. He said he'd tell Fred, the managing editor. By Tuesday, I was down to one testicle, swaddled, and anxiously undergoing follow-up tests to find if my cancer had spread. That day, I got an unexpected email:

Asher,

Hi, it's Lance Armstrong here. I heard yesterday from
Fred about your situation and wanted to drop you a
line. Of course I'm well familiar with what you're going
through and am hoping for the best. I'm confident you'll
be fine! I would recommend (after the pathology report)
a trip to see either Dr. Larry Einhorn (Indiana) or Dr.
Craig Nichols (Portland) to get a truly world-class second
opinion. I can help with this if need be. I'm in Johannes-
burg, SA right now but home on the weekend so just let
me know if I can help. Otherwise, hang tough and keep
livin' strong!

Best,
LA

It felt like an unexpected reprieve, a get-out-of-school note in the middle of the week. This was early 2006, remember, when Lance, an Austin resident, was at the height of his popularity—when he was still engaged to Sheryl Crow and the official winner of seven Tours de France. I was like the cliché of the sick kid visited by the famous baseball player who promises to hit a home run. I soon learned my cancer—a particularly aggressive strain—had spread to my lymph system.

I called Einhorn's office to see if he'd be free. Not for about six weeks, unfortunately, his nurse told me. That seemed an eternity: I didn't want to sit on my hands while a nasty disease was rapidly working its way through my body. So I wrote Lance back, asking for his advice. Within an hour, this answer:

> Asher,
>
> No worries at all. I have already emailed Dr. Einhorn and told him you'd be calling. If there are any problems then let me know.
>
> Take care,
> L

Einhorn's office could fit me in the following week, the nurse said this time.

As with so many things in my life, I was lucky to know someone who happened to know someone. But my sense is that in thousands of ways large and small, Lance had given some hope or help to a person unexpectedly facing their mortality. The following October, already rid of cancer, I participated in my first Livestrong Challenge, an easygoing race that raises money to fight the disease. Besides the siblings, parents, friends, and children of cancer victims, there were the survivors and the patients: bald kids hanging out in the backs of pedicabs; wigged women jogging with their families; and people like me, just happy to be on my feet. As I crossed the finish line with Rebecca, I was given a survivor's yellow rose. The race gave us a sense of traction, of fight and engagement, against an often random-seeming disease. We felt relief, yes, but we were moved also by the realization that many of those around us were at less fortunate points in their story than I was, and that this meant that what we, as a group, were celebrating was not so much a lucky escape—a happy ending—as something more essential to the character of each one of us, and less dependent on providence.

In the lead-up to Lance's confession to Oprah, I wrote a personal essay, a defense, in the newspaper. Some commenters had appeared gleeful about this latest turn of events involving Lance and doping allegations. "From Hero to Zero," read one online headline. Not surprisingly, I had a different take. The sanctimonious epithet that followed Lance around was "arrogant," as if that's not a trait commonly found in the best athletes. Arrogance is part of what makes great athletes great: They play, they jump, they cycle with the expectation of greatness. Early in Michael Jordan's career, his coach supposedly scolded him that "There's no I in 'team.'" Michael's response: "But there is in 'win.'" (Charles Austin, too, was a sort of peacock. Once, as I huffed-and-puffed through some sit-ups, he began yammering on his Bluetooth. "All y'all were on all that stuff and I still beat y'all. It don't matter what you were on—I still won. I still won." "Who's he talking to?" I asked Terrell as I rolled over. "There's no telling," said Terrell with a smile. "He's like this every day.") It takes arrogance, surely, to take on an opponent as blindly formidable as cancer. It takes arrogance to parlay inborn physical prowess into the idea that one man can make a difference against such an overwhelming Goliath.

That arrogance had a dark side, of course. In January 2006, Armstrong said in sworn, now cringe-worthy, testimony: "I would never beat my wife, and I never took performance-enhancing drugs." None of the testimony at the time turned up eyewitnesses to Armstrong's doping, and SCA paid Armstrong $7.5 million (which the company is now trying to get back from Armstrong). But in 2008, with scientists who had testified for SCA trying to throw doubt on his journal article, Coyle was forced to acknowledge an error in his calculations. His most famous academic article was now caught up in the controversy. Still, even with Armstrong's 2012 confession, Coyle felt a loyalty to Lance, despite the admissions of cheating and worse. You are left with these two very different men: Lance, with a reckless, ruthless drug-taking regime, designed to boost his natural talent, and Coyle, dutifully jabbing himself only to record, for the sake of science, his solid but far more modest capabilities. Coyle was Daedalus, the master craftsman, seeking to improve performance through science; Lance, a phenomenal talent, one who justifiably became a hero because of the work and money he devoted to fighting cancer, was an Icarus-like protégé who, in a fit of hubris, had soared, with man-made aids, too close to the sun. In his grief, Daedalus busied himself with his work, determined to unravel, with human know-how, the sorts of mysteries that bedeviled the ancients. The same could have been said for Coyle.

Now here I was, in the exercise machine myself. "Go, go, go!" shouted Coyle—echoing Polly in New York City—as I struggled to rotate the wheel from a standstill. Finally, after a solid second of muscle recruitment, I pushed through. My result: 5.3 watts per pound, which put me just below the average power of a female college rower—and significantly lower than a men's college basketball player. With just a couple of months to go, I was grounded enough, I had learned, that there was no chance I would melt near the sun.

13

So, Can White Men Jump?

46 days left: *I catch myself lingering in the mirror a couple of beats longer than necessary—the vanity that comes with a newly shaped body. I can squat 265 pounds. "Bring the heavy" is the expression some of the other weightlifters use—and that's as much as I can muster. But to coax explosiveness out of my muscles, I concentrate now on the weighted squat jumps—three sets of 20, at 75 pounds—and the usual sets of lunges, leg presses, calf raises, and upper-body work. This saves my right knee some aching: It prefers quick action with lighter weight, like a lottery winner choosing yearly payments over a lump sum.*

When, some years ago, my brother-in-law Ben dunked in a game on West Fourth Street, in Manhattan's Greenwich Village, a passing businessman pressed his face to the chain-link fence to cheer: "Nice to see one of us doing that!" When people say white men can't jump, they aren't being completely literal, of course. Dunking is a stand-in for style, and the stereotype is that white guys, shuffling about in their Asics and khakis and polo shirts, are incapable of coolness. When the businessman cheered Ben on, he was appealing to something beyond tribal identification and loyalty—though there was, ridiculously, plenty of that, too. The comment was pregnant with envy. The sociologist Gena Caponi-Tabery has argued that the slam dunk, like a piece of jazz, is a form of expression that embodies characteristics of the African-American aesthetic: improvisation, virtuosity, and defiance. Whether it's a riff by Miles Davis or hang time by Michael Jordan, the moments that linger in our minds involve some exceptional, stylish individual skill that combines these traits. If you can dunk, in other words, you can at least lay a tenuous claim to coolness.

But the businessman's comment also tapped into the conventional wisdom that some things, genetically speaking, are closed off to most of us. "Look, man, you can listen to Jimi but you can't hear him," Sidney Deane, played by the smooth-as-silk Wesley Snipes, tells the countrified Billy Hoyle, an especially honky-looking Woody Harrelson, in *White Men Can't Jump*. Rebecca, who can quote the movie backward and forward, reminds me that Billy does eventually dunk ("But it looked fake," she says with a light shake of the head)—and we can all hear Jimi, even if we don't get the lyrics quite right. But when word got around that I was trying to dunk, smiles spread across faces: "White men can't jump," friends would say, as if it was my first time hearing that. "You look like a desk guy," one acquaintance told me doubtfully, and I couldn't help wondering if it was my hue to which he was alluding.

At the 1996 meeting of the British Association for the Advancement of Sport, the distinguished Roger Bannister, the first breaker of the four-minute mile, decided to weigh in on differences between blacks and whites—a subject outside his expertise. Blacks had a "natural advantage," in running, Sir Roger claimed, as if an entire race of people were born with sneakers on. He wasn't alone in idiotically, and reprehensibly, holding forth on the topic: "The black is a better athlete to begin with because he's been bred to be that way," sports commentator Jimmy "The Greek" Snyder observed in a notorious 1988 television interview. "This goes back all the way to the Civil War when during the slave trade . . . the slave owner would breed his big black to his big woman so that he could have a big black kid." Jack Nicklaus was once asked why there were so few blacks among the elite ranks of golfers. "They have different muscles that react in different ways," he said.

Sports commentators have long played on the specious differences: White athletes are hardworking, black athletes are amazing physical specimens; whites are smart, blacks are talented; whites are team players, blacks are showboaters. In his 2000 book *Taboo: Why Black Athletes Dominate Sports and Why We're Afraid to Talk About It*, a kind of sports entry in the race and culture wars of the period, Jon Entine—a journalist, not a scientist—delivered what he called a "biocultural" explanation that "biological factors specific to populations can exaggerate the impact of small but critical anatomical differences." He observed that athletes of African ancestry hold every major running record, from the 100-meter dash to the marathon. "Genes may not determine who are the world's best runners, but they do circumscribe possibility," he wrote.

While he had his supporters, including his sometime collaborator Tom Brokaw, Entine quickly drew a varied and expert set of critics. "If professional excellence or over-representation could be regarded as evidence for genetic superiority," Jonathan Marks, an anthropologist at Berkeley, wrote in the *New York Times*, "there would be strong implications for Jewish comedy genes and Irish policeman genes." The science about the relationship between genetics and performance was still too vague, he claimed. "We could document consistent differences in physical features, acts and accomplishments until the Second Coming and be entirely wrong in thinking they're genetically based." Essayist Jim Holt wickedly opened a review of Entine's book with the conclusions about sporting and genetic makeup from an earlier period:

> It is pretty obvious that certain racial and ethnic groups are naturally gifted at playing certain sports. Take basketball. That's a Jewish sport. So, at any rate, people thought in the 1930s. After all, the star captain of the original New York Celtics, Nat Holman, was Jewish, as were four of the starters among St. John's famed "wonder five," who ruled college basketball in the late '20s. Jews were believed to have a genetic edge, being endowed by nature with superior balance, greater speed and sharper eyes—not to mention, in the words of one sportswriter, a "scheming mind" and "flashy trickiness."

(If only my own flashy trickiness, along with my usurious ways and abiding appetite for virgin Christian blood, translated into dunkability. Scheming gets you only so high.)

Still, the issue whether African-Americans were naturally superior in some sports has flummoxed many, including Arthur Ashe. "Damn it," Ashe once said. "My heart says 'no,' but my head says 'yes.' Sociology can't explain it. I want to hear from the scientists. Until I see some numbers [to the contrary], I have to believe that we blacks have something that gives us an edge."

So what does scientific research tell us about genetics, race, and athletic ability? Or, put less delicately, can a white guy like me dunk?

On the face of it, the answer is obviously yes. White guys do it in high school gyms every day across the country. But the very question—asking whether a white guy can dunk—fundamentally misunderstands why some of us appear naturally better at things than others. "'Race' is the wrong term," for thinking about these issues, Harvard evolutionary biologist Dan Lieberman tells me. Muscle development has nothing to do with the color of one's skin. But scientists have noted common characteristics among "biological populations," or people from different parts of the world, says Lieberman. In other words, we ought to separate race from our ancestral geography, just as we might distinguish a man's clothes from his politics. Just because someone appears to be black (or white or brown or whatever) doesn't mean we can draw any conclusions about her physical abilities. Knowing where her ancestors came from can sometimes offer hints. The genealogy of some top-notch long-distance runners, for instance, has been traced to a particular Kenyan valley; some sprinters' family trees find their roots in specific parts of West Africa. "You have more individuals in those populations better built for power than endurance," says Lieberman.

Just because you or your forebears hail from Côte d'Ivoire obviously does not mean you are going to be a particularly good sprinter. But, says Lieberman, certain pockets of the West African population are more likely than others to have what's known as the speed gene, alpha-actinin-3, or ACTN3, a protein found exclusively in fast-twitch muscles. Fast-twitch muscles produce more force more quickly than slow-twitch muscles; squid use theirs to shoot out tentacles to catch prey, we use ours to sprint after a bus. The downside is that they're quicker to tire than their slow-twitch counterparts, which can withstand repeated contractions over a long period of time without lactic acid buildup; these are the muscles that mussels use to cling long and hard to slippery, briny rock as they're relentlessly beaten about by waves. For similar reasons, the turtle beats the hare.

•

Michael Bárány weighed only 92 pounds when the Americans liberated Buchenwald in April 1945. "I was just of bone," he would tell an interviewer from the Shoah Foundation a half-century later. Raised the son of a well-to-do farmer outside Budapest, he was denied admission to university for his Jewishness and as young man had found work as a mechanic in a Hungarian army factory. In late 1944 he and other Jewish mechanics were directed to a train depot and told that they were being dispatched for a military operation. They were in fact being sent to a death camp. They were given a two-day supply of food to endure a 21-day journey in a boxcar—some of the ninety men, including Bárány, resorted to drinking their own urine. "Hunger, smell, and disease transformed the cattle wagon into hell," he later wrote. By the time they were dragged out of the boxcar, one of the men had died and two of them could not walk. The others were made to march through the Buchenwald gates, above them the taunting slogan "*Jedem das Seine*," or, roughly, "Everyone gets what he deserves." It was Christmas Day, 1944.

After the liberation, Bárány, 23 years old, weakened by tuberculosis, remained nearly a month at the camp to recuperate. He then made his way back to Hungary, where he learned his parents had been gassed in Auschwitz. He had grown up an Orthodox Jew, but God had been wrung out by the Holocaust. "I tell you when I lost God," he told the Shoah interviewer. "I lost God first when I heard my parents saying God would help. And when I became a scientist, a life scientist, then I couldn't find God anymore." By September 1945, determined to establish a new life for himself, he enrolled in university.

Bárány specialized in medicine. He met Kate Fóti, a fellow student, when she came into his dormitory to get first aid after she cut her finger slicing some bread. Within months they were married. She would become his lifelong scientific collaborator. After winning his medical degree, Bárány turned to biochemistry, drawn to the lab of a Hungarian Nobel Prize winner. He began to study the lives of muscles, and how and why they contract. I like to think of this man, whose formative years were spent suffering at the hands of the most inhumane ideologues, concentrating his meticulous study on what seems to me to be our inner humanity: the source of both our grace and our power, the tools that control everything from the gesture with which we soothe a child to the vigor with which we dunk a basketball.

But he faced a second kind of prejudice after getting his doctorate in 1956. He found himself unable to teach under Communist rule because of his father's prewar landholder status—no matter that his family land had been wrested away by the Fascists. He was deemed a capitalist. Lacking opportunity, the Báránys plotted to flee the country. The border was mined, watched, and wired. "So escape was not absolutely simple," Bárány later observed.

At one point the couple paid a former secret police officer to help them escape, but he was arrested. In another scheme, they were to be smuggled aboard a coal barge departing Budapest for Vienna, but the Danube froze and the ship was jammed. They finally engaged a guide to help them cross the woodsy border into Yugoslavia. He steered them to a spot where a cow had stepped on a land mine and been blown up, thus creating a zone for safe passage. The family—Michael, a scientist on the verge of a breakthrough; Kate, seven months' pregnant now; and George, their 22-month-old toddler—trudged through ten miles of snow, zigzagging through the night to dodge the border guards. Bárány carried two suitcases, one filled with food, the other with diplomas and lab notebooks. Family lore had it that at the end of this harrowing trek, the boy, who had been sedated with a sleeping pill and carried for the first few miles before being made to walk, told his parents upon their arrival in Yugoslavia *"Jól sétáltunk"*—"That was a nice walk."*

In 1960, after stints in Israel and Germany, the family finally settled in the United States. Living in the fringes of Queens, the Báránys left their house so early for work that by the time their boys awoke they had only a governess to greet them. They were dutiful sons: their mother made them a poster of 17 good reasons to do push-ups, and the boys did push-ups routinely upon getting out of bed each morning. (Not far away lived my grandparents, themselves refugees from Vienna.) The Báránys' destination each morning was Manhattan's Institute for Muscle Disease, where they worked on teams trying to cure muscular dystrophy, a disease of progressive weakness, often first seen in children, that hampers locomotion. The Báránys were "inadvertent comparative biochemists," they would later say, examining the muscle contractions in rabbits, chickens, and frogs. It was in New York, doing these experiments, that Mi-

---

* The Bárány story resonates with me. A few years ago, my family reconnected with my grandmother's cousin, a man much younger than she, named Milan. He turned up in Ottawa, where he was working as an engineer. Milan's very existence had long been a sort of rumor. The story was that his parents and his older sister had survived the war in hiding, each separate from the others. Milan was born after the war, once the family reunited. As with the Báránys, the Hungarian property belonging to Milan's family was seized, never to be returned. Sensing little in the way of opportunity, they decided to flee—to Austria, where some relatives lived. But the crossing attempt went very wrong. They were caught by border guards, possibly because their guide was a turncoat. Milan's parents and his sister were shot to death. He himself would have been killed had his mother not shielded him. As it was, he was shot in the back. When the family never showed up in Austria, the news made it all the way to my grandmother, in Queens, another in a line of tragedies. No one knew for sure what had become of the boy, though there was word that he had survived and been adopted. About five years ago, my aunt tracked him down. He was as pleased to get to know us as we were to get to know him—he had known virtually none of his biological family. Milan, twinkly-eyed, round-faced, intellectually minded, and his wife, Laeora, are now fixtures at our Passover seder. They were present, in fact, at my grandmother's last seder in 2013, and I think her death struck him as profoundly as it did us.

chael Bárány made his most acclaimed discovery, of a relationship between the speed of muscle contraction and the breaking down of energy-bearing molecules inside the muscle. The finding, one that characterized the difference between "fast-twitch" and "slow-twitch" muscles, was laid out in a 1967 *Journal of General Physiology* article that has now been cited more than 1,700 times. Nearly a half-century later, this was the difference I was hoping to navigate in my own training as I steered my body to be fast-twitch heavy.

In the mid-1970s, the Báránys found work in the biochemistry department at the University of Illinois at Chicago. Their partnership led to many joint publications—72 full papers and a dozen book chapters. "To my parents, science was a safe haven because it could not be taken away, as so much else had been," George would later say. "Science was logical and predictable in a world where they had experienced so much suffering based on the arbitrary prejudices and madness of human beings." George and his brother, Francis, both prodigies (one graduated from Stuyvesant High School at 16 and went straight to graduate school), became scientists in their own right. At Chicago, Kate and Michael Bárány walked to work together every day; they became known on campus as "the professors who held hands."

Kate died in June 2011; Michael, six weeks later.

•

To understand the relationship between the speed gene and muscle performance, you need to understand something about how muscles work. Muscles are the body's most bounteous type of tissue, and arguably the most trainable. They are like sets of Russian nesting dolls, with one biological structure nestled within another. Every muscle fiber, as long as a foot but thinner than a strand of hair, is composed of long tubes called myofibrils. These, in turn, are made up of thin and thick protein filaments running in parallel. Every muscle fiber, be it fast- or slow-twitch, has both sets of proteins. A muscle contracts when the thick set of protein filaments grabs the thin set; because the thin set of filaments is attached to the end of a muscle fiber component, the entire fiber is thus shortened.

A biology researcher once explained it this way in the magazine *Nature*: Picture yourself standing between two hefty bookcases. These bookcases, being in the basement of a massive archive, are on rails so they can be easily moved. It's your job, standing between them, to bring the bookcases together, but you're limited to using only your arms and ropes hanging down from each bookcase. (Your arms are the thick filaments; the ropes are the thin filaments; the bookcases are the ends of the muscle fiber.) Standing between the bookcases, you pull on the two ropes—one per arm—that are tied tight to each bookcase. In a repetitive fashion, you yank each rope toward you, re-grasp it, and then yank again. Eventually, as you make your way through the length of each rope, the bookcases move together.

This effort—of pulling bookcases together in a stuffy basement—is sweaty business. It requires energy, especially if you want to tug those ropes (that is, contract your muscles) quickly—this was the relationship that Bárány had illuminated. Flush with mitochondria—the little power centers within each cell that use oxygen gas to convert sugar into chemical energy molecules—slow-twitch muscles can perform low-intensity exercises for long stretches. Thus, the Ethiopian distance runner Haile Gebrselassie, at one point the holder of 27 world records, has muscles that are likely beefsteak-red, juiced with oxygenated blood. Fast-twitch muscles, on the other hand, are short on mitochondria. Were we to peel away Usain Bolt's skin like an onion, we would likely see a paler, whiter set of muscles—something like the color of pork. When a burst of energy is required, rather than undertake the plodding oxidation method of their slow-twitch counterparts (also known as aerobic metabolism), these muscles quickly break down their large reserves of stored sugar via anaerobic metabolism, which is much faster but also much less efficient. Scientists think humans working at their maximum speeds can rely on non-oxidized forms of energy conversion (or anaerobic metabolism) for about 30-some seconds before lactic acid develops and muscle fatigue sets in. By then the fast-twitch muscles, contracting roughly twice as fast as their slow-twitch counterparts, have done their job, like overdrive on a sports car.

Each of us is born with a certain ratio of fast- and slow-twitch fibers. That raw number does not change. But when we work out, our muscles clearly get bigger. That's due to an increase in the number of those protein filaments within each fiber, not an increase in the number of fibers. So while the number of fibers is not changing, individual fibers are gaining protein and thus gaining in size. We can focus that growth within a certain fiber type: We can increase the size and strength of fast-twitch fibers or slow-twitch fibers, depending on the exercises and training we do (in my case, fast-twitch muscle training). This will not change a slow-twitch fiber into a fast-twitch fiber. Suppose your muscles are 50 percent of each. Training will not change that percentage, but it can make the fast-twitch fibers larger, so an entire muscle group like the biceps is now more than 50 percent fast-twitch by mass or bulk. That would make you more fast-twitchy even though the raw number of fibers of either type has not changed.

Back to the speed gene: Scientists are still unsure exactly what ACTN3 contributes to the performance of fast-twitch muscles—and most think it pales as a determinant of athletic ability compared to training time and competitive will. But they suspect

ACTN3 influences the speed with which we grab energy-bearing molecules to do work in explosive activity. Every one of us has two copies of the ACTN3 gene—or its variant, a mutant that coincides with a deficiency in the actin-building protein associated with fast-twitch muscles. (We have two copies because we inherit one copy of the gene or its variant from each of our parents.) So it follows that each of us has one of three possible pair combinations of these two genes: two ACTN3 genes, suggesting we're predisposed for power events; two mutant genes, suggesting we're predisposed for endurance challenges, like long-distance running; or one of each, suggesting we might be decent—but not necessarily great—at both sorts of activities.

Researchers have consistently found that power athletes have at least one, if not two, copies of ACTN3. In a 2008 paper in the journal *Genetics and Microbiology*, a team of Greek researchers reported that the ACTN3 gene showed up in the top power-oriented athletes considerably more often than in a representative random sample of the Greek population. Even among non-professionals the presence of the gene correlates with what might be called your twitchiness: Another Greek study—the Greeks, haunted by their long-faded Olympic glory, evidently obsess over the science of track and field—found that boys who lacked a copy of the ACTN3 gene were significantly slower in a 40-meter sprint. In a nutshell, the gene "has a predictive value if you're fast-twitch dominant or not," Lieberman, the Harvard biologist, told me.

Still, a surfeit of fast-twitch or slow-twitch muscles does not mean you will be a great sprinter or a great long-distance runner. Training, courage, and tenacity play bigger roles. But Lieberman tried to disabuse me of my leaping hope. "If you never have had the great dominance of a fast-twitch athlete, you'll never be a great fast-twitch athlete," he said. And then, twisting the dagger with what I could swear was a bit of glee, he told me: "I'm willing to bet that you'll never become a great jumper. The older you get, the harder it is to change that muscle fiber composition."

•

Genes alone don't make someone a good basketball player. Writing of his fellow African-Americans, John Edgar Wideman once observed that "our stories, songs, dreams, dances, social forms, style of walk, talk, dressing, cooking, sport, our heroes and heroines provide a record . . . so distinctive and abiding that its origins in culture have been misconstrued as rooted in biology."

Families build identities in similar ways. Keen to suggest bookishness runs in our blood, my own family fancifully traces its lineage to a fifth-century BC scholar—BC! Take that, you *Mayflower*-descended motherfuckers!—named Ezra the Scribe. Twenty-five hundred years ago, with many of the Israelites stranded in Babylonian exile, the Persian king granted Ezra, a renowned scholar eager to return his people to their homeland, passage to Jerusalem to teach the laws of God. When he arrived, after a six-month journey, he found that left-behind Jewish men had been marrying non-Jewish women—Samaritans, actually. (The good ones didn't come around until centuries later, materializing when people needed help with their flat tires.) Ezra tore his garments in despair before ordering and enforcing the dissolution of such sinful wedlocks. He then declaimed the Torah to assembled Jews, and the people and the chief rabbis swore to keep themselves separate from gentiles. I take an amused, skeptical pleasure in my rabbinical roots, but I was disappointed, having scoured the Bible and the Midrash, the Talmud, and other sources, to find no evidence that Ezra had any jumping abilities. Nothing.

We like to gaze upon the more recent, greener branches of our

175

family trees to explain our current generation's talents and predilections. I tell the story that I chose journalism in part because it was my dad's first career—but it was an inheritance separate from the blue eyes he bestowed upon me. As a swashbuckling college reporter in the late 1950s, he wrote dispatches from Russia and revolutionary Cuba. My father had himself been enthralled by the ultimately grim stories about his Viennese great-uncle, Gabor Engelsman, who ran the *Sonn-und-Montag Zeitung*, a newspaper with a reputation for reportage and satire discomfiting to the Austrian authorities and hardly reverential toward the Germans. With the Anschluss, the annexation of Austria by the Third Reich that precipitated my grandparents' flight to the U.S., Gabor knew the Nazis would arrest him, and he jumped from an office window to his death. As a boy, the fates of my forebears were relayed to me, of course, as reminders of the Holocaust. But they were also meant to describe lives that might have been, and one of those was another in a dynasty of quick-thinking journalists.

But is there anything in my genetic code that suggests a talent for asking questions or sharing stories? Sounding out the echo of the parent in the kid is an old and confounding pastime. Maybe one can tell whether a toddler has the mouth of the mother. But the gait of the father—is that inherited, or learned? To what extent was Allex Austin's jumping ability learned through the intimate, lifelong observation of his father, Charles, and to what extent was it genetically conferred upon him? I asked Terry Todd, the former champion weightlifter who presides over the Lutcher Stark Center, an Austin museum and library dedicated to the study of what the human body can do, whether strength ran in his family. Visitors to the center are greeted by a cast of the colossal Farnese Hercules—vulgar, carnivorous, and mounted on a faux-marble, slow-spinning turntable. We were seated at the large oval mahogany table in the center's library, a life-size oil painting of Arnold Schwarzenegger, in Speedo and mid-flex, looming behind us. Todd, still a massive, broad-chested man at age 76, looks as if he could overturn all sorts of furniture, even this table, if he were so inclined. The question, straightforward as it was, is one he had long pondered himself. When he was about nine years old, he told me, out fishing with his grandfather, a "mallet-handed" 5'10" Texas rancher, the old man picked up a native pecan, an oblong nut guarded by a shell so tightly constructed it requires a special nutcracker to break, and crushed it between his thumb and forefinger. "Bud," he told Todd, "few men and no boys can do that." Todd himself would become a world champion power-lifter, but try as he might, he was never able to duplicate the feat his grandfather pulled off. It's a story he also tells in "Philosophical and Practical Considerations for a 'Strongest Man' Contest," an essay about designing obstacle courses challenging enough for outsized humans. His grandfather "broke quite a few more such pecans for me, and each time he did it he chuckled, especially since it was a feat I was never able to match no matter how large I became or how hard I tried," he writes. "How was it that a man in late middle age who had done no systematic training could be as strong in any part of his body as a much younger man who stood four inches taller and outweighed him by over a hundred pounds?"

•

I was gathering spit in my mouth, hovering over the little tube I had ordered from 23andMe, a genetic testing company named for the number of pairs of chromosomes coded with genetic information that are littered throughout our bodies. I was hoping to learn more about the genetic makeup of my muscles. Rebecca was sitting next to me, directions in hand, eyes averted.

"You don't mind terribly, right?" I said to her as I gathered a bit more saliva on my tongue. For years following my cancer diagnosis, Rebecca treated the slightest symptoms with obsessive anxiety. She worried that headaches might be signs of brain tumors; that a cramping foot was an early signal of MS; that a sore throat suggested esophageal cancer. The idea of a test that promised something like medical clairvoyance—my future health history laid out for me—both drew and appalled her.

Seven years out from my cancer, I still make annual visits to my oncologist. Routine as they are—a draw of blood; a listen, by stethoscope, to my lungs; a gentle squeeze of my remaining testicle; a quick massage of my lymph nodes; an X-ray, to see whether any new tumors are forming about my chest—they fill Rebecca with a profound anxiety. Though she's a confirmed atheist, after my initial chemotherapy treatment she bought me a jade Chinese talisman of an abstracted unicorn. (Einhorn, the name of my oncologist, means unicorn in German.) She thought I could wear it for luck, not so much as a safeguard of good health as a symbol of it. One evening, I foolishly, accidentally, broke it, dropping it into our porcelain sink as I prepared for bed. It cracked in jagged halves. Of course the accident meant little, other than the breaking of a gift from my dearest, but the split medallion gained for her, despite herself, a half-real cosmic significance. With Rebecca fretting about the smallest possible ache, I had to check myself before telling her about back pains or headaches and stifle coughs—all, oddly enough, possible signs of a return of the cancer. But I myself didn't get anxious about these doctor visits. I was, in a sense, a lightbulb, and Rebecca the moth.

I had embarked on this dunking project to see how far my flesh and bones could take me, to know about the frontiers of my capability; Rebecca, a self-described homebody—an eater of plain bagels, a vanilla-ice-cream girl—shunned frontiers. She didn't want to know what the Fates had planned for me. But by dint of scientific breakthroughs in the mapping out of the genome, we now can have a pretty good sense of our natural potential for everything from dunking a basketball to going blind. Here, suddenly, in a way that overturned all that was sacred in Western literature, from Adam onward, free will would be reduced to a website that presented me with the cold data of my essence. "Know thyself," Socrates commanded. But surely the ancient Greeks had something very different in mind; *Oedipus Rex* is warning enough of the risks of prophecy. The mysteries of my future, including the sort of natural death I might endure, would be distilled to a game of percentages, as if I were a deck of cards. Is that sort of certainty really a desirable thing? Rebecca had wondered, as I breezily plonked down the hundred bucks to be dealt my hand.

"You don't mind, right?" I asked as a kind of afterthought. I had grown so accustomed to her anxieties that I could be cavalier about them, and even, at times, dismissive.

"Let's not talk about it," Rebecca said.

"Because of the cancer stuff?" I managed to warble as I spat, for the fourth time, into the test tube.

"Because it's gross," she said.

It was hard to disagree: My DNA was now slithering down a bit of plastic I gently held between my thumb and forefinger.

It was time, I had decided, to figure out whether my genetic makeup had put me at a competitive disadvantage when it comes to dunking. It's easy to discover whether you (or your child) has the so-called sports gene: In 2008, Atlas Sports Genetics began selling, for $149, a screening test for variants of the gene ACTN3. Since then at least a half-dozen other companies have gotten in on the act. Federal regulators are dubious. Not long after I took this test, the Food and Drug Administration sent a letter excoriating 23andMe for failing to prove up its testing and notification procedures.

"The commercial tests are very reliable in the sense of telling you which genetic variants you have," Vishy Iyer, a molecular geneticist and microbiologist at the University of Texas, told me. "But in terms of actually predicting something about your success in one sport versus another, they're currently unproven at best, or worthless at worst, in my opinion." He estimated I had an 80 percent chance of having at least one copy of the speed gene. "Many people have this fatalistic sense that our lives are determined by our genetic code," Iyer continued. "But with sports that's really unknowable. It's hard to get at." Iyer, a tall, thin man—a marathoner, with elbows slightly akimbo, as if he were ready at any second to start racing—took me seriously enough to get out of his office chair and chart, on his whiteboard, the relationship between genes and muscle mass. But as he filled the board with jottings and arrows, the air busy with the moist squeaks and faintly chemical scent emanating from his clutch of red, green, and black markers, he turned to tell me that in the final analysis the ACTN3 gene is only a small component of physical activity. The question of athletic excellence is far more complex than one simple gene or its variant. We started having the sort of conversation that straddles philosophy and science, about magical twins who were identical in every respect except one: One possessed two copies of the gene, the other no copies. The first twin might have a marginally better ability to dunk than the second. I was sorry Rebecca wasn't with me; she gets excited about twins, mostly because she and her twin are extremely close. "Twinpathy," they call their mind-meldedness. If Rebecca spots twins, infant or adult, when we're out for a stroll, she pokes me surreptitiously. Sometimes I think she wants to give a little wave of acknowledgment, the way bus drivers do as they pass each other in opposite directions.

•

Eight weeks after putting the test tube in a biomedical bag and popping it in the mail, I got the results via email, of course.

Rebecca, for her part, didn't want to know any of it, especially the parts predicting my medical future. "Haven't you submitted yourself to enough tests? Haven't you had to wait for enough results? Haven't you been poked and prodded enough?" she asked plaintively.

"There was no poking; I only drooled into a tube."

I argued that the results were probably predictable. My father had endured prostate cancer, for example, and my maternal grandmother had suffered through Alzheimer's—that must leave me at greater risk for both. Oddly enough, she didn't find my way of looking at it all that comforting. As a concession, I decided to forbear from clicking open most of the health results, confining myself to information relevant to my jumping genealogy. It would be the virtual equivalent of covering your eyes at a scary movie, peeking only now and then through a web of fingers to catch fascinating bits of backstory.

Between you and me, a tiny part of me had hoped to find out that I was partly black. But no, I learned from the backlit laptop in my living room, I am fully Ashkenazic, as plainly white as one of those long fluorescent office bulbs. On my mother's side, my genes suggested I belong to a branch of humans who trace their lineage to a quartet of women who lived around 700 AD, according to the genetics company. Those women, who settled in what is now Germany's Rhine Valley, were Jews, and their descendants now number in the millions. On my father's side, my genetic makeup is similarly Western European. (I'm in the 87th percentile of Neanderthalishness, I'm proud to say; the genetics company tells me that 3 percent of my genetic makeup can be traced to our distant hairy relatives, as if I didn't know that by just taking off my shirt.)

What I really wanted to know was whether, genetically speaking, I was swimming with or against the current. Was Vishy Iyer correct that I was almost surely in the genetic middle, in the muddle of humans with one, but not two, copies of the speed gene?

No, as it turned out. The genetic test turned up neither of the copies of the speed gene. I am apparently in the narrow minority of only about 18 percent of humans who lack even a single copy. My Neanderthal qualities notwithstanding, one way of looking at this result is that, physically speaking, I'm highly evolved; through a long chain of natural selection, some humans have cast off their sprinting abilities, which are far less needed than they were hundreds of thousands of years ago, when our arboreal selves still sprung about for their dinner. I'm better suited, I was learning, to the long game, the persistent stalking of prey over great distances that became our M.O. as we descended to the plains. Another, less happy thought: Despite my height and long arms, the genetics were always stacked against my dunking endeavor. Thanks a lot, Mom and Dad.

July 15, 41 days left: *Michael Jordan's outstretched hand measured 11.375 inches from tip of thumb to tip of pinkie. Mine's about half that. One way of thinking about this: I'm adorable. But if your hand is too small, you have to hold the ball gingerly, painstakingly, like a child handling a spoon in an Easter-egg race. There's nothing I can do about hand size, obviously, but I can improve tensile strength. So now I'm undertaking a few fingertip push-ups. And while watching television my left hand dangles down, my fingertips clamping a basketball like a spider perched atop a pumpkin. I try to keep it squeezed for 30 seconds, or until my fingers ache.*

Progress report: With a month or so left, on an unusually cool summer night, cool enough that I thought I could palm the basketball, even after warm-ups left my skin moist, I decided to try to dunk. The outdoor court at the nearby Boys & Girls Club was deserted. I was grateful. A live oak, a fairy-tale tree with low, sprawling limbs, easy for climbing, gave me some cover from any passersby. No one would catch my embarrassing efforts. I started a run-up at the rim and felt the ball slip out of my hand. I tried again. Same thing. Again with the run-up, and again the ball slipped out. Okay, take it easy. I pulled back and shot the ball a few times, blew on my hand, as if it had just handled an unexpectedly hot pot, wiped it off on my shirt. And grabbed the ball, solidly. I walked to the three-point line, squinted at the basket like it was an outlaw in an old Western, squeezed the ball at my hip, and drove forward. I took off and brought my left hand high into the air, as if I were an airborne Statue of Liberty—and pushed the ball down into the hoop, the fingers of my left hand sliding against the cold orange metal as the ball rattled down. "Holy shit!" I shouted, to no one in particular. I had dunked a full-size basketball on a rim that adults used to play basketball. Except it wasn't quite a 10-foot rim. Still, I was thrilled. Like a little boy, I ran home the block and a half to tell Rebecca the news.

A few days later, on a Sunday afternoon, I went back out with Rebecca to get the dunk on video. I wanted to prove it to her as much as to anyone else. A little blond boy broke off a walk with his father to press his face to the fence as I tried—and failed. "Honey, keep up," the father said to the boy as he continued walking. "But Dad, he's going to dunk." You're right, wee one, I thought. And here I go . . . and I missed again, the ball slipping out of my hand like I was playing water polo. The boy padded away. I wondered, for a brief minute, whether I had imagined my triumph of the other night. But I banished the thoughts and convinced myself, Charles Austin–style, that I had already dunked on these baskets. (After all, I had.)

And so it happened, in this year of our Lord 2013, that I dunked a basketball, before my lovely wife and on video. And I let out a roar. Be it noted, once more, sotto voce, that it was on a rim a couple of inches short of regulation. But I was pumped for the full-on dunk with just a month to go.

14

It's Gotta Be the Shoes

July 23, 33 days to go: *I saw Hugh Jackman on Let-terman tonight, plugging his new Wolverine movie. Man looks ripped. He reported—and this made me very sad—that he didn't eat any fruit during his months of working out. Feeling guilty, I solemnly pledge, from here on out, no fruit for me. Not just fresh fruit; no dried fruit, either. No raisins in my oat-meal, no grapes as a snack. No apples. No banana-frozen cherry-nonfat milk licuados, whipped together in my blender. Nothing. I had grown up in a family that regularly grazed on things fruity, as if scurvy was right around the corner. Fruit was so cherished that gnawed-at apple cores were not deposited in the garbage but stuffed in the refrigerator between the Sunday-brunch lox and years-old marmalade, there to wait patiently for some mysterious future use.*

In autumn 1988 a newly minted Reebok executive named Mark Goldston flew from Boston to Michigan on a top-secret sneaker mission, code-named Dr. Detroit. Goldston wanted to see for himself what he hoped would be the "holy grail" of athletic foot-wear—a pair of shoes that, in defiance of the sorts of natural limits we all face, would actually make its wearer jump higher or run faster. He had been chief marketing officer at Reebok only a few months, a 33-year-old advertising maven plucked from the toiletry and fra-grance firm Fabergé, where he pitched antiperspirants and hairspray for men, to revitalize the stagnant sneaker company. Before Fabergé, he had run the marketing shop at Revlon for four years, and now, in a way, he was being asked to do a makeover job at Reebok. The com-pany, named for the graceful South African gazelle, had a proud, if musty, history. Descended from a late nineteenth-century English footwear purveyor, the company still insisted that a Union Jack grace every shoe. Reebok had made its mark in the States during the 1970s aerobics craze, and, again, in 1982, with the introduction of a woman's running shoe, a novelty at the time. But as the decade heated up, aerobics and running were taking a backseat to basket-ball, a sport with new celebrities eager to expand their reach and income through endorsements. In 1988, Reebok saw its first drop in earnings, about 20 percent. The company, lacking a high-profile high-top, was being lapped by its competitors. Nike had its Air Jor-dan; Adidas had its Attitude, favored by Patrick Ewing and other big men; and Converse, the old, classy standby, had its Weapon, worn still by Magic Johnson and Larry Bird. Reebook's only basketball shoe was the BB4600, which was affectionately known inside the company as a shoe you'd wash your car in.

Chicago-raised, Goldston had always been a brash overachiever, willing his 5'11" frame to dunk at least a volleyball in high school. "I was always two inches away," he says ruefully. (I felt his pain.) The narrowness of that difference, between ordinariness and mar-velousness (or, as they say in the makeup industry, between looking good and looking great), fascinated him. The slogan of his child-hood shoe brand, PF Flyers, rang in his head: "Run your fastest, jump your highest." Was there a shoe that could get you to run even faster, jump even higher, than your natural capability?

That was the hope with Dr. Detroit. Reebok had hired a "podiatrist type of guy" at the University of Michigan to design a sneaker with springs inside—"a tiny trampoline" is how Goldston describes it—that promised to instantly improve the hops of its custodian. This was the prototype that Goldston dragged himself to Michigan to try. He still played ball now and then, and he wondered whether these shoes would be what he needed to overcome those two inches.

They were duds. After running and jumping in the clunky shoes for a half hour, Mark Goldston's feet fell asleep. The size of the springs made them too bulky. Plus, he wasn't sure he was jumping any higher—partly because they were so big. Still, he took the shoes back to Massachusetts to conduct future tests. At home he entrusted the sneakers to his dependable quality control experts, his identical twin sons, Adam and Ryan. Little more than toddlers, they were old enough, already, to compete at basketball—they competed at everything, really, a pair of talented left-handers who hated to lose. They were convinced they would one day dunk, and as they slipped their tiny feet into the massive Dr. Detroiters, their eyes grew wide, enchanted by the idea of a shoe with special powers.

•

About the time Adam and Ryan Goldston were toying with the Dr. Detroit moonboots in their Massachusetts manse, I was a gawky prepubescent trying not to show up late for my Amsterdam Avenue school bus. As an 11-year-old in New York City, my favorite shirt was a baggy gray tee that hung past my butt emblazoned with a photo of Patrick Ewing, the great Knicks center, rising up to slap away an opponent's shot. "Return to Sender" was stamped across it. This was as much as we could hope for, advertising-wise, from Patrick, marginalized by corporate America for his Jamaican accent and workmanlike ways. We New Yorkers were haunted by a more charismatic dunker, a contemporary of Ewing's: Michael Jordan.

Not only did his Bulls repeatedly shoo our Knicks from the playoffs, but Jordan's image was omnipresent—on sneakers, cologne, T-shirts and, of course, television. Jordan put his popularity down to the "ability to do things that other people can't do but want to do and they can do only through you." "They watch you do it, then they think that they can do it," he told *Esquire* magazine in 1990 for an article titled "Michael Jordan Leaps the Great Divide." Advertisers glommed on to this logic, too. In their genius, they held out the possibility that we mere humans, even Knicks fans, could approach Jordan's abilities, if only we bought the right merch. "It's gotta be the shoes," Mars Blackmon, the alter ego of Spike Lee, famously declared in a series of Nike advertisements in the late 1980s and early 1990s, as Michael dunked in the background. (In the pre-*Seinfeld* era, the Nike ads were arguably the best thing on television. They were clever and stylish and they respected the intelligence of both the viewer and the mostly African-American stars who made the commercials what they were.) Good fun, of course. No one really thought Michael Jordan's talent had all that much to do with his sneaks. Even Michael said, "No, Mars." But the message got through: If you wanted to "be like Mike"—as the jingle went in the Gatorade commercial—you ought to wear his shoes and drink his drinks. Rebecca, being the Michael Jordan fan she is, can still sing the lyrics: "Sometimes I dream / That he is me / You've got to see that's how I dream to be / I dream I move, I dream I groove / Like Mike / If I could be like Mike." The song is no more about dunking than the Mars Blackmon commercials are. They're about upward mobility, yes, but about a particularly American version of gaining fortune and fame. That's what it means to be like Mike, who—along with, perhaps, Bill Cosby's Dr. Cliff Huxtable—was one of the first African-Americans to truly be a star for black America and white America equally. It's what Mark Goldston was promising to consumers with his sneakers, in what was the heyday of sports footwear. For white kids, "movin' and groovin' like Mike" meant borrowing some black cool—in that way, just wearing the sneaks were a stand-in for the dunk. At $100 a pair, they were also an early version, for both blacks and whites, of bling—a sign of flushness. And it's why, in a way, the sneakers themselves, especially the white-and-red Air Jordans, kept squeaky clean, became a status symbol of the streets, of wealth and power, something that teenagers famously killed each other over. (In a coda matching our times, the Air Jordan and its sneaker contemporaries are now offered their own shelves and accent lighting in SoHo's museum-like, hipster-curated boutiques.)

With my dunking deadline nearing, and desperate to grab any spare half inch I could find, I began wondering, just as Mark Goldston once did: Could something as simple as a change in sneakers make me more like Mike? I had just learned that I was genetically deficient in the fast-twitch-muscle department. How easy would it be to buy my way out of my natural physical inheritance?

Advertisers have long peddled potential. Whether they're selling weight-loss programs or penis enlargement, companies aim to convince consumers they can remake their bodies. The only thing many baby boomers remember of their otherwise forgettable 1950s comic books were the Charles Atlas advertisements on the back covers that depicted a scene at the beach: "Hey, Skinny . . . Yer ribs are showing!" a beefy bully shouts at an adolescent standing next to a bikini-clad girl. Two panels later, after the boy has subscribed to the Atlas regimen, he returns to the beach to kick the bully's ass. "Oh, Joe! You *are* a real he-man after all," the girl tells him. In the bottom half of the ad appeared a picture of Charles Atlas, flexing a biceps.

People used to laugh at my skinny 97-pound body. I was ashamed to strip for sports or for a swim. Girls made fun of me behind my back. Then I discovered my body-building system, "Dynamic Tension." It made me such a complete specimen of manhood that I hold the title "The World's Most Perfectly Developed Man."

It was no coincidence that the insides of these comics were dedicated to wondrous tales of transformations, from weaklings to superheroes. There was Robert Grayson, the diminutive son of a Jewish scientist who flees Earth with his family during the rise of the Nazis, only to return as Marvel Boy, with Uranian powers to fight baddies. And there was Steve Rogers, the scrawny offspring of Irish immigrants, who, injected with a serum by the U.S. military, becomes the unstoppable physical specimen known as Captain America. These transformations often involved little hard work—a radioactive spider here, a super-serum there. With the Atlas ads pictured on the backs of comics for nearly 50 years, the implicit lesson for generations of lightweight introverts was that they, too, could undergo a transformation.

●

On a sunny, unseasonably warm February day in 1989 in Atlanta, Mark Goldston, accompanied by thumping music, bounded onto a stage at the Sporting Goods Manufacturing Association meeting, the world's big sneaker trade show, to tell convention-goers about the invention that would secure his name in the annals of sneaker history. Even with the Dr. Detroit project scrapped, Goldston was convinced he could apply the principles of marketing to sneakers by marrying the technology craze to the high-top. That year, *Back to the Future, Part II* ruled the cineplex and Apple had recently released its first portable computer. Keen to the zeitgeist, footwear companies started using "visible technology" in their sneakers—little plastic portals that purported to give a window onto the performance-changing Asics gel and the Nike air. Goldston had already patented the Energy Return System, a design of cheap plastic tubes that promised to release energy with each footstrike; he had overseen the development of a sneaker whose color could be changed with the insertion of a cartridge; and he had devised a campaign around honeycomb padding found in the seats of the space shuttle.

"I wanted to make something the Jetsons would wear," he says.

The sneaker Goldston would present that day in Atlanta took its inspiration from a pair of his wife's Scandinavian buckleless ski boots that had a built-in device that sucked in air to keep the boot snug. Goldston sketched out the shoe with Reebok designers, ordering that a self-contained pump in the shape of a mini basketball be placed on the outside of the tongue. The prototype used rubber bladders made by a Massachusetts blood-pressure-cuff company.

"The chairman wanted to call it ActivAir," Goldston says with a snort. "I was like, 'They have a whole company based on 'air'"—Nike—"'and it'd just sound like we're copying them.'

"I suggested 'The Pump,' and someone says, 'That's what women call their high-heeled shoes,' and I said, 'Do you really think fifteen-year-old boys will know or care what women call their heels?'"

The shoe industry began buzzing about the Reebok Pump and this young sneaker wizard. Advertisements with Dominique Wilkins followed, leaving the vague impression that the shoe would increase powers of levitation. Dee Brown, of the Boston Celtics, would famously stoop down during the 1991 NBA Slam Dunk Contest to pump up his Reeboks before attempting one of his winning jams.

The price was set at $170, making it the most expensive mass-produced shoe on the market. "If we're going to do something, we've got to shock the system," Goldston remembers telling his colleagues. "For Chrissakes, we've got a Union Jack for a logo. What is that?" The high price for The Pump was also a sales gimmick to get consumers to settle happily for at least a $75 shoe. "This was going to be like the Mercedes SL in the showroom, the one that sells all the E-Classes," says Goldston, using an analogy that suggests how well he's done for himself. But the Pump was a phenomenon in its own right. "Guess what? We sold the SLs at E-Class volumes."

The Pump became a status symbol, a suburban kid's attempt at seeming urban and black and rich at the same time. "Pump it up!" was the unwittingly masturbatory cri de coeur shouted on blacktops across the country. But no one took the scientific claims with anything other than guffaws; the Pump carried all the seriousness of a walking elementary school science fair experiment.

Goldston, who had become known, somewhat derisively, as the High Priest of High Tech, soon left Reebok, in a squabble over sneaker performance. He moved the family to California and took over L.A. Gear, where his major accomplishment, in another techno-twist, was installing lights in the heels of sneakers. Eventually he would abandon shoes altogether to do Internet startup work. But the thing that haunted him—as much as anything haunted a guy who seemed to float from one corner office to another—was the Dr. Detroit project, and his inability to find his holy grail.

The Dr. Detroit misadventure had stuck with his kids, too. Adam and Ryan, the twins already shooting balls at age four, had become bona fide athletes. They had worked ceaselessly at improving their hops through school while leading their Van Nuys prep school team to an end-of-season title. Sharp, handsome, self-confident—they look like they stepped out of a GQ photo shoot—they even made it as walk-ons on the USC basketball team. But they knew playing basketball wasn't in their long-term future, and by 2009, still enthralled with the idea of a technological route to the rim that would appeal to the average player, they convinced their father to join them in a new version of the Dr. Detroit project. They were all of 22 years old. After sketching out more than a hundred designs—the sneaker exec versions of Leonardo da Vincis—the family settled on one with eight springs in the forefoot of the shoe, in a nest of material similar to an egg carton, housed, like an Oreo cookie, between two plates.

The kids knew—they had learned from the best—that at the end of the day the shoe business was really a marketing game. And so they went about selling these cheap-to-produce, made-in-China high-tops with their kitchen-table technology as some kind of space-age scientific breakthrough. As if to confirm the value of the springs, they set a high price point—$300, just as their father had done a generation earlier for The Pump.

They decided to sell only online, to avoid the distribution costs that come with brick-and-mortars. They came up with a name, cleared with lawyers, inspired by NASA's Jet Propulsion Laboratories: Athletic Propulsion Laboratories, meant to evoke both jerseys and lab coats. And in a bit of marketing moxie, they patented their "Load 'N Launch technology," basically a fancy term for the spring near the front of the sneaker. The sneakers, they would explain on their website, are "uniquely designed to capture the maximum amount of energy through the compression caused by exerting pressure on the forefoot and then releasing the energy through the propulsion and liftoff stage to increase vertical leap." The shoes would "provide an instant advantage to virtually everyone who wears the product and engages in jumping on a basketball court." They commissioned studies that purportedly showed the sneakers added as many as three inches to a person's vertical—just by dint of lacing them on. It was the jumping equivalent of liposuction, or Botox—an instant upgrade.

The strategy paid off big-time when the NBA, in October 2010, announced it would ban the shoes because "under league rules, players may not wear any shoe during a game that creates an undue competitive advantage." The Associated Press, ESPN, Yahoo! and Sports Illustrated picked up the news. The APL website crashed. The Goldston clan was ecstatic: The NBA ban appeared to offer authentication of their claims. The Goldstons trumpeted the decision in news releases of their own. Visitors to the Athletic Propulsion Labs website—the ones who could log on—found a picture of the shoe stamped with the words "Banned by the NBA." "It's the ultimate validation," one of the twins told a Los Angeles television news reporter.

What's bad enough for the NBA is good enough for me: Giddy at the prospect of an easy couple of inches, I ordered a pair of the Concept 2 APL sneakers. This was my best chance, I thought, to fix my genetic foibles. I would essentially buy myself some talent. When the plain cardboard box arrived several weeks later, I greedily opened it and gazed at my chariots of fire—black, with bolts of electric blue and neon green and yellow. But I couldn't bring myself to put them on. I was terrified that they wouldn't work, that, in the end, they would be ordinary, earthbound high-tops. We never want to dash our own fantasies. Rebecca has declared that she wouldn't want to go to Hawaii because it cannot possibly be as awesome as she's built it up in her mind. Or maybe it's like being in high school and never mustering the courage to ask out your crush. It's the terror of what would happen if you confess your feelings. You can't be turned down if you never ask. You keep the fantasy alive. And so I put the sneakers back in their box, in a corner of the living room; if they didn't contain magical powers, they could keep that secret to themselves awhile longer.

It ended up taking me a few weeks to pull together the nerve to test my thunder-bolted new jumping shoes. I decided, finally, that the shoes wouldn't be of any help if they weren't tied to my feet. So early one summer evening, I laced them up courtside and warmed up, getting comfortable in my sneakers and loosening up my body. By now I knew these rims intimately, like a ship captain knows the feel of her tiller. I knew just how much rim I could grab, and with what authority. And, finally, after thousands of jumps, I knew myself. I was hoping for the dunker's equivalent of Dorothy's ruby slippers. I'd only have to stomp on the ground three times, and then—wham!—the shoes would take me home. I expected some of the "dramatic sensation of lift" APL promised in its promotional material as I flew up to the rim. I gave it a whirl, first with no ball, then with one.

Nothing. Not. A. Thing. They didn't help at all. I found myself reaching no higher on the rim than I had before. I didn't feel any spring. I tried them again each night that week. The company says that "certain athletes have found that the more they wear Athletic Propulsion Labs shoes and understand what it [*sic*] can instantly do for them, the more benefit they seem to derive." Forget for a moment that that makes no sense—what kind of instant benefit takes time?—I was finding that compared to leaping in my plain old Nike high-tops I was getting no higher. Unlike Dorothy's, my new slippers were not taking me where I wished to go. I felt like a fool, having put store in a fantasy. I had asked the girl out, and she had indeed turned me down, like a bedspread.

I wondered about the NBA ban that started the Athletic Propulsion Labs craze in the first place. The company's website has a section titled "Science of Jumpology" that pulses with phrases like "the resultant effect" and "integrated response curve," but had the league actually tested the APL shoes? Tim Frank, the league's head of basketball communications, told me that no player or team ever actually came to the league with a request to use the shoes. League officials had met with Athletic Propulsion Labs at the company's request; company officials made claims about how their shoes could improve jumping ability; the league went ahead and banned the sneaker.

Athletic Propulsion Labs asserts that "independent testing" at an unnamed U.S. university verified the company's data: According to the company's website, the university researchers found that eleven of the dozen participants "jumped higher instantly" in the APL shoes than in conventional sneakers. Could it be that the shoes were magical, and it was I who was defective? I asked John Porcari, a professor of exercise physiology at the University of Wisconsin–La Crosse who tests advertising claims by exercise equipment manufacturers, for advice. He told me to ask APL for a copy of its scientific study. Examine how it was designed, he said: Did the researchers randomize the sequence in which jumpers tested the sneakers? Did the wearers jump first in conventional sneakers, giving them time to warm up before trying the APL ones? "You can design a test that gives you certain results," Porcari said.

He added: "The company should have nothing to hide. They should want to give it to you."

•

"I'm in world-class shape," Mark Goldston told me, when I reached him, mid-workout, at his Beverly Hills home. He himself had once argued that Americans work out as part of a "narcissistic" urge to look good—not exactly profound, but at least a bit of candor from a sneaker executive. Now 58, Goldston works out nearly three hours a day. "I've got eight percent body fat!" he said, with some giddiness.

He was charmed by my dunking ambition, and immediately nostalgic for the great dunkers of his youth. He was a big Dr. J fan. And he told me about the search for his holy grail. I asked him what made him so sure the Athletic Propulsion Labs sneakers could help their wearers jump higher. He recited the information on the website about the stunning test results.

"Could I could see the raw data for myself?" I asked.

"Sorry, that's not possible," he said. "The university made us sign a nondisclosure agreement." The university—a "leading West Coast university" was as specific as he was willing to be—"told us they can't take part in 'commercialization' of a product—they can't be seen to endorse a product," he continued.

"Well, how about you at least send me a copy of the report with the names of the university and researchers redacted?"

"Can't do that, either. It'd violate the terms of the nondisclosure agreement."

As a newspaper reporter, I've run across instances of companies forcing university researchers to sign nondisclosure agreements—everyone from tobacco companies to solar energy corporations that want to suppress data or present it in a favorable way—but never the other way around. Mark Goldston at least conceded that the NBA probably didn't undertake its own tests, instead relying on claims the company itself made based on research it had commissioned—except, of course, the company wasn't willing to release the actual research results.

Before we got off the phone, I shot him one last question: So, if these shoes had been around when you were 20 or so, could you have graduated from a volleyball to a basketball? Could you have closed that two-inch gap?

Either because he's a true believer or a very savvy marketing man, he didn't skip a beat: "There's zero question in my mind that I would have been able to dunk."

15

Jumping Secrets of the Ninjas

July 25, 31 days left: *My fruit fast lasted exactly 35 hours. Who doesn't have an appetite for apples? I refuse to apologize for eating a banana. Besides: a piece in the* Journal of the American Medical Association, *brought to me by my trusty nutritionist, Rebecca, showed that increased fruit consumption lowers body weight, because all that fiber makes us feel full and improves our metabolism. Shred that, Wolverine!*

In the months following my cancer diagnosis, several friends privately revealed to me problems with their testicles, partly to express empathy, which I didn't really need, partly to get advice, which I couldn't really give. These revelations were usually in the form of phone calls, sometimes from people I hadn't heard from in years. The conversations started with some pleasantries and then, invariably, the tone would turn more confidential. "Listen, I want to ask you about something," a friend began. "What did the swelling feel like?" He was sure one of his testicles was bigger than the other. Another thought he detected a bulge after the family dog had leapt onto his lap. Following my treatment, I wrote an account of my cancer experience in the newspaper. One reader emailed about his testicular pain after sex. "Cancer?" he asked.

I was reminded of this as word got around about my dunk project. Friends of friends approached me, as if I were a therapist, to tell me about how badly they had once wanted to dunk and how, even now, they wondered whether they ever could. One, a guy like me in his early thirties, told me animatedly that he had long dreamt of slamming the ball home, and videotaping it to prove his manhood. As a comeback to just about anything, he could whip out the imaginary tape: "I could just say, 'Oh, yeah? Well, take a look at this!'" Now, stuck with a table-waiting job, he told me that the couple of hours he spends working out are the best of his week. "I live for that time," he said. He's a married man, and I found this admission a little sad. Another friend of a friend, a Mexican-American who grew up poor, to parents who were cotton pickers, in one of the *colonias* of rural South Texas, told me about his own brief dunking past. As a senior in high school he was thin as a cornstalk—6'1" and 130 pounds ("when wet," he says)—but athletic as hell. He played on the school tennis team and had to do so many jumping drills to improve his agility that he was eventually able to dunk. He told me dunking was a better feeling than winning at tennis. A single moment of absolute thrill, and I wondered whether I would ever feel it.

Given how much of my life was now given over to it, jumping became an obvious thing to talk about at parties. "Tell them about your dunking!" an Austin hostess would say. Then, somewhat timidly at first—I know, I know, this sounds silly, I would mumble— and then, with more confidence, I would hold forth. "The science of jumping?" someone would inevitably say, while giving a little hop. "You must, you just must, study Baryshnikov," said another, delicate in her movements. "I insist."

At one such shindig, a woman in her 40s was just explaining to me that her husband, a young, handsome doctor, had spent that very morning at the Mormon temple, baptizing the dead, when the host popped over and asked how the dunking was coming.

I said it was coming along, but I was uncertain about my chances.

"Have you tried jumping over hemp?" the Mormon woman chimed in.

"Excuse me?" I said.

"I mean, a hemp plant," she said, as if that would make things much clearer.

She saw that I was totally confused.

"Suzuki says that to jump higher, the ninjas used to jump over a hemp plant each day."

"Who is Suzuki?"

"He's the famous violinist," she said.

Why would a violinist care about jumping?

Suzuki, she explained, had written a book about nurturing talent, and in it he recalled the habit of the ninjas of jumping, daily, over a slow-growing hemp plant. As it grew, imperceptibly, their leaps steadily, quietly, grew higher with it.

The story put me in mind of my 12-year-old, book-obsessed nephew, Dovid, and his advice about jumping higher: Jump over a stack of books; each day add a book to your stack. This was the Jewish version of the ninja tactic, apparently.

"Can I legally even get a hemp plant?" I wondered aloud.

And then the Mormon doctor, clean-cut, with beautifully trimmed black hair, wearing a white polo shirt and khaki shorts, leaned in: "I hear that if you ask where you can pick up some basil at a head shop, they can help you out."

16

Carrying Capacity

22 days left: *Battling knee soreness, I stick to light weights and short sprints. After a long stretch and a plyometrics set, I do a dozen sprints of 25 meters apiece. At the gym: hamstring curls, calf raises, leg extensions, and 85-pound squat jumps. I retire to the sauna—my sweet relief—and crease open my novel, Moshin Hamid's* How to Get Filthy Rich in Rising Asia, *careful not to sweat all over it. A tattooed man, shirtless with sneakers and shorts, asks me for the book's best advice. Fair enough.*

In early August, only a few weeks from the finish line, I headed back to New York for a final examination.

After I took off my shirt, Polly, now an old friend, gestured up and down at my body: "Just look at him," she said warmly. I might have blushed. "You look like a new man." I was thinner for sure, and more powerful. My body fat was at just 8 percent, down from nearly 21 percent at the start of my year. On the Wingate test, I now clocked a peak power output of 932 watts, a solid 20-percent increase since the first time I oafishly stumbled into Polly and Jamie's lab. I was approaching the quick-twitch power of—wait for it—the 1999 Israeli national basketball team. My people! My standing vertical, meanwhile, had gone from 18 inches to 21 inches—still merely mediocre, but an overall 17-percent increase in my jumping ability. There's still some spring in these bones. And, even better, my leaping vertical—which harnessed the momentum of a quick sprint into upward force as I leapt off my right foot—was about 27 inches. Now I could get my full hand above the rim, a far cry from when I strained just to touch it. Yet still too short—if only by an inch—to dunk a full-sized basketball on a full-sized rim.

I had a couple of weeks still left—but my legs were starting to break down.

•

In my regular life I write about environmental issues for the daily newspaper in Austin. The environmentalists I interview like to talk about "carrying capacity," or how much life a given plot of land can naturally sustain, in terms of habitat, food, and water. They argue that it's unnatural to build suburbs atop environmentally sensitive areas. Take the increasingly fragmented Texas Hill Country, for instance, home to endangered songbirds and salamanders—and now a fast-growing suburb of Austin. Once upon a time it was one of the poorest areas of the nation, a land of rocky soil and inhospitable drought, until a young congressman named Lyndon Baines Johnson, a native son, steered money in the 1930s to the building of enormous dam projects that brought along electricity. With its thin soil and scarcity of water, environmentalists argue, the Hill Country is not meant for the crush of roads, swimming pools, homes, baseball fields and all the other things that materialize in a twenty-first-century burg.

The human body has its own carrying capacity. Even the greatest athletes among us are circumscribed not only by forces like gravity but also the muscle mass, bone density, and connective tissue of their bodies. The pinnacle of the 2013 season for Matt Harvey, the 24-year-old pitching ace of the New York Mets, arrived on August 7, when he threw his first complete-game shutout, baffling the Colorado Rockies with his torrid fastball. Already he had started for the National League in the All-Star Game and had hurled two near no-hitters. But not long after that game against the Rockies, his elbow began to bother him. The diagnosis: He had torn a ligament. He had thrown so hard that his arm just broke. Harvey, the flamethrower, would miss the entire 2014 season following ulnar-collateral ligament surgery, better known as Tommy John surgery, for the pitcher who originally underwent the procedure in the 1970s. The surgery is remarkably successful, and horrifically common—as many as one-third of starting pitchers now undergo Tommy John surgery to keep their careers going.

Glenn Fleisig, research director at the Birmingham, Alabama–based American Sports Medicine Institute and an expert in pitching mechanics, tells me that we've reached the limit on how fast humans can throw a ball. You can improve the mechanics, the conditioning, the nutrition, the muscle mass of an athlete. But at some point the tendons and ligaments are too weak to support the superhuman whiplash movements; the body just doesn't have the carrying capacity. "You get a good athlete, not me or you, some minor leaguer, he seems athletic or studly and we signed him from New Jersey or Venezuela or wherever, he throws 85—and we want him to throw 95," Fleisig tells me, and at this point I'm wondering how he can tell, just by hearing my voice, that I'm not a good athlete, that I'm not studly. "The limiting factor is not how strong he makes his muscles, it's what your ligaments and tendons can take. These guys are not coming in with broken bones or torn muscles—it's the torn ligaments that are the problem. The ligaments and tendons are taking maximum loads of over a hundred pitches in a single outing."

Fleisig tells me that while pitchers have generally maxed out, sprinters and jumpers have lots more room for improvement. He makes the observation that his colleague, James Andrews, a famed orthopedic surgeon (he's that guy every famous athlete visits after an injury), "does not have sprinters who blew out knees and hips" hanging around his waiting room.

There were obvious ways I could have extended my own carrying capacity. "Steroids" is a dirty, unctuous word, one that conjures episodes of surreptitious injections into teammates' butts. For me, they once served a nobler purpose, fortifying my body as chemotherapy toxins broke it down. But I eschewed pharmaceutical assistance this time around. Beyond my physical squeamishness, I worried I would somehow change my very character, fall prey to an awful fit of 'roid rage. I didn't want to hulk out that badly.

So what is the natural carrying capacity of athletes? How fast, given maximum conditioning and talent, can we run or swim? How high can we jump? How far can we throw a ball? Absent a natural or artificial physical evolution, would the athletes of the future perform any better than they do now? Some have tried to get at the answers. A Chicago doctor has predicted that one day we could see a 7'2" superathlete with freakishly long arms and an eye-opening 51" vertical *dunk on a 14'5" rim*. In a 2010 article in the journal *Mathematics and Sports*, a professor at Bloomsburg University in Pennsylvania calculated with 90-percent certainty that the speed limit for the 100-meter dash lies at 9.4 seconds. Making such calculations is slippery business: The professor conceded that before Bolt's world-record dash of 9.58 seconds in 2009, he would have predicted

a fastest-possible time of 9.62 seconds. Past performance is no guar-
antee of future results.

I wondered if, boiled down, we were like the mathematical limits
we had learned about in high school, with that arrow pointing to a
finish line it never quite reached. A life spent approaching a goal but
never quite arriving.

●

After getting the look-over from Polly, I met with Steve Doty, the
Hospital for Special Surgery researcher who had first taken an interest in my project, at the hospital coffee shop. Doctors in scrubs
padded around, pouring milk into little bowls of cereal.

I told him of my worry that I was coming up just short.

"You could always try limb lengthening," he said.

Medieval punishment wasn't such a bad idea, I supposed, if I
could dunk at the end of it. Everything has a silver lining. "You
mean the rack?" I laughed.

"No, we actually do limb lengthening here," he told me, "for patients who have had birth defects or serious accidents."

There are instances of cosmetic leg lengthening—men at 5'3", say,
who want to be closer to 5'9". These typically involve the intentional
breaking of bones as doctors then pull fragments apart to encourage
growth. Doty wasn't seriously suggesting I go that route, but the fact
that cosmetic lengthening even exists illuminates the extraordinary
lengths people are willing to go to shrug off their natural limits.
Bone-breaking struck me as a clear violation of the one promise I
had made myself, to do no medical-grade harm. More than that, it
felt like a violation of nature. I like my bones just the way they are.
Besides, my year was fast coming to an end.

Not long after I left New York, an encouraging report came out of Chicago, one that should cheer would-be dunkers everywhere: Derrick Rose, the Bulls point guard who had sat out the entire previous season during his recovery from major ligament surgery, and who had set his own return mark to when he could dunk off his injured left leg, announced that he had added an extraordinary five inches to his vertical. "I think I jump higher," he told reporters. "I think coming into the league I was at thirty-seven [inches vertical jump] and they tested my vertical at [a training facility]. I'm probably at, like, a forty-two [inches], so I'm jumping a little bit higher." I felt like his year and my year were weirdly aligned. If Derrick Rose, already renowned for his explosiveness, could squeeze out a few more inches of jumping talent following a devastating injury, couldn't we all? And Josh Scoggins, one of the young men I had met at the Hill Country basketball camp, who had sworn off jumping because of growing pains, wrote me. In pregame warm-ups, he had completed his first-ever dunk. "Everyone screamed, and then I did it a second consecutive time," he said. "It felt awesome."

PART III
the TOUCHDOWN

17

Allez Hop!

In Austin there grows a type of agave called a century plant. For nearly all its life this bluish, smooth-leaved agave appears inert and bloodless. It grows very slowly. And then, seemingly overnight, it sends a bright green flowering stalk dozens of feet in the air. The stalk, straight as Gary Cooper, resembles a gigantic uncooked spear of asparagus, one so thick a grown man can't get both his hands around it. You can find the century plants, so called because the stalks shoot up after decades of dormancy, across the deserts of the Southwest, lending an otherworldly feel to the dry plains.

I think of the sprouting of the century stalk as the great shout of the otherwise-silent cactus. The plant goes, suddenly, from a period to an exclamation point. But it also marks the end of the book, as it were, for the plant. It is exhausted by the reproductive effort. Beneath the stalk that has waved high for months, the leaves of the cactus wilt and gray. Sickly, they flop to the ground and then, finally, the stalk itself crumples into a leathery decay.

The effort it took to dunk, the months of preparation, the daily pushing open of the doors at the gym to lift weights, the sprawling on the ground yet again to do still more push-ups, the counting up to 15 with near-nauseating repetition—fifteen reps of biceps curls, fifteen reps of triceps presses, fifteen fingertip push-ups—to tease out that potential that I trusted lurked invisibly within me, was coming to a close. I was expecting a spectacular final flourishing before my project's end. Eventually, my muscles would melt, my belly folds return, my age assert itself. Already my legs were tiring, my knee sounding the alarm that time was against me.

Astonishingly, with only days to go, it remained a mystery to me whether I could dunk a full-sized ball on a full-height rim. That I was now this close, the equivalent of an underdog with a shot to win the game with the clock ticking to zero, was itself amazing to me. Once, a couple of months into my project, I had suffered dreams in which coaches, unhappy with my lack of dunking ability, shipped me out of the country—to France, peculiarly enough. Now I found myself sliding into sleep with visions of slam dunks. Still, in plain daylight I knew that I couldn't simply run up and dunk the ball on a 10-foot rim, as I had at the ever-so-slightly-short Boys & Girls Club rim. Bearing the ball with me during my run-up, I couldn't re-create the angle I would need to push the ball down. My hands were just too small, my tensile strength too weak, to both palm the ball and sharply swing my arms in such a way that I could steam ahead.

My solution to this problem: the alley-oop dunk. Etymologists trace the name for one of the most electrifying moves in all of sports to the cry of French acrobats as they were about to leap into a trick. "*Allez hop!*" they would cry, in their presumably impeccable silent-*h* French accents. (It means "Off you go!" or "Let's go!") The term "alley-oop" then surfaced in the late 1950s, used by the NFL's 49ers to describe the lob play to their springy 6'3" receiver R. C. Owens, a former college basketball player. The quarterback would loft the ball high into the air, trusting Owens to soar and snatch it. Within a few years the term took hold in the NBA, popularized by the late Los Angeles Lakers announcer Chick Hearn, to describe the exhilarating play in which one player lobs to another for the midair catch-and-dunk. Executed properly, the alley-oop scans like a poem: beginning as a long-syllabled hang-in-the-moment, floating finesse pass in the proximity of the rim; ending, the ball plucked in midair, with a short, accented phrase, the bursting dunk. Deciding to go with an alley-oop dunk to finish my year required particular audacity on my part. The alley-oop takes coordination, concentration, and timing, after all. It's a don't-try-this-at-home sort of dunk. These would be extra, final hurdles in my dunking year. But it also had tactical benefits: Attempting an alley-oop meant that I would not have to carry the ball rimward, thus freeing my arms for their pistonlike work. And if I could pull it off, an alley-oop would be an emphatic way to nail my first—and perhaps last—ever real dunk. All I had to do, easy-sounding enough, was palm the ball in the air and snap it through the hoop.

●

The day of the dunk was a Monday in late August. The following afternoon Rebecca and I would board a plane for Istanbul, for Nathaniel's wedding. Dunking, Texas, stretching, and dieting would recede from my mind at 550 miles per hour. This was the last chance.

I had taken it easy the previous few days, track-star-style, tapering off my workouts to rest my leg muscles before their big liftoff. As I undertook my standard warm-ups at Charles's gym—the skips, the high-knees, the bounding, and so on—I felt fresh and loose. My nerves melted away. Charles dribbled the ball by the basket and I readied myself at half-court, running through, in my mind's eye, the keys to getting up.

A small audience gathered: Rebecca, there for moral support—and to take video of what might be an epic moment; Charles's assistant, Terrell; a half-dozen or so 12-year-olds, all aspiring dunk-ers, who took a break from their basketball game to watch; and the few customers who happened to be hanging out at Charles's gym that morning. It was high noon in Texas, a faceoff between me, at half-court, and my nemesis, the rim. The gym had grown quiet, except for some murmurs, and everyone was staring at me. All the work had come down to this, a final series of dunk attempts. "You ready, man?" Charles asked. And on that first jump, I exploded up, swatting the ball toward the back iron and then hanging, for a second, on the rim. It gave a serious shudder as I dropped to the ground. *Yes, I was ready.* Everyone was watching, egging me on, now, hollering. The kids took their shoes off, sat near the basket, and beat the floor with their sneakers during my run-ups. I was getting higher up than I ever really thought possible, but I was not quite putting the dunk down.

If you had stepped inside the air-conditioned gym from the big-sky brightness and throbbing Texas heat outside, you would have seen a man in modest, solitary, periodic flight. He's sporting not a cape but a curious costume: a crimson gym T-shirt (increas-ingly darkened by sweat) with the slogan "Gold Is Our Standard" stretched across the back (trumpeting Charles Austin's gold-medal-winning history); a frankly quite ugly pair of Athletic Propulsion Laboratories sneakers, just in case they decided to start matching their promises; and navy-blue super-lightweight short shorts—practically a high-tech loincloth—meant for runners (no extra-long baller's shorts for him). Nearly beneath the basket, you would have seen a second man, strong and loose-limbed, emanating compe-tence, bouncing the ball with two hands and tossing it underhand into the air, like a basketball ref letting go a jump ball. You would hear Katy Perry's "Teenage Dream" playing on the sound system in the background, partly drowned out by the clanking of weights that echoed about the cavernous space, softening the squeaks that squeezed off the parquet floor as the man pawed at the half-court circle with his sneakers like a batter stepping up to the plate. You could have watched his first attempt. His second. His tenth.

Again, again, the ball would leave my hand, only to rattle against the inside back lip of the rim, and out. I could not get my arm high enough for the right angle to slam it true, as if I were a carpenter incapable of swinging his hammer up to bash the nail-head home. Fifteen tries. I glanced down at the palm of my left hand to find the joints of my middle and forefinger bloodied from scraping against the rim. I reached back to that cold winter morning a year and a half earlier when I had met Tyler in Central Park: He had warned me of the bruising and the blisters, tracing a line of callus across his fingers. I had secretly doubted I would ever get high enough on the rim for such a thing to happen to me. And now, on the last day of my great dunking experiment, my fingers were bleeding. I was shocked and pleased. They appeared, these blood blisters, like dunking stigmata, mystical evidence of the real pain I was feeling.

"I'll put money on it," said Terrell. "Show me something!"

Twenty minutes of this passed. Some of the kids, inspired by my exertions, were now lined up to dunk on a seven-foot plastic hoop in the corner. Talking trash through giggles, they were playing grown-up. I envied them; they made it look easy.

I collected myself. I sipped some water from a fountain. I shook out my legs. I readied myself again. Again I flew up to the basket, my wrist above the rim, my palm kissing the ball.

●

Deep in the New Mexico desert, in a razor-wired compound guarded by the military, I glimpsed the future of jumping. Buffeted by the wind, hills of red rocks scrabbled around, the federal Robotic Vchicle Range felt like an outpost on Mars. It was home to a few double-wides, a mini-warehouse, and a communications tower—a modest collection of buildings that form the initial testing grounds of some of the world's most advanced robots. It was in that warehouse, off a high shelf, that Jon Salton, a government engineer, pulled down an orange hard-plated suitcase, one that a Romanian might have traveled with in 1976. Salton laid it on a workbench, snapped it open, and pulled out something that looked like a metal mini-skateboard with puffed-up wheels. This little machine was why I had made the trip to the high desert: This was the Hopper 2.5, a six-pound proof-of-concept vehicle that promised to remake jumping from a high-spirited expression to a serious-minded tool of warfare.

In the late 1990s, the Defense Advanced Research Projects Agency, the arm of the Defense Department that funds newfangled technology like algae-based fighter-jet fuel, began supporting efforts to build a robot capable of leaping at least 20 feet while reliably maneuvering across complicated terrain autonomously. Historically, wheel size has determined mobility: the bigger the wheel, the higher the clearance, which is what you have with those ridiculous-sized monster trucks. The hopper robot, all six pounds of it, was meant to break the paradigm. Now you can have a small vehicle that simply hops over an obstacle. "Where there's a flat spot—and there are many of those in this world—it's a whole lot easier to use the old-fashioned wheel," Salton's boss told me. "Still, there are places this can go that you can't take a tank."

Researchers at the Robotic Vehicle Range won the nod from DARPA to solve the challenge. Part of Sandia National Laboratories, the range was carved out of the vast Kirtland Air Force Base, a left-behind, Cold War–era kind of place, still dedicated to designing nuclear weapons. A Sandia researcher named Rush Robinett, an avid trout fisherman, got the idea for the machine's early navigation style as he was out catching grasshoppers to use as bait.

"I noticed they jump around in a random fashion, hit the ground in an arbitrary orientation, right themselves, and jump again," he said. "I said to myself: 'I can make a robot do that.'" These prototypes were the size and shape of grapefruits and coffee cans, with low centers of gravity. They were durable, but they fell over, or barely made it off the ground, as scientists tinkered with the right fuel mix.

The original plan was to prepare the hoppers for interplanetary exploration, to dispatch them in every direction from a Mars lander. But 9/11 changed all that, with DARPA desperate to find robots that could do reconnaissance in the Afghanistan theater. Over about eight years, a dozen people worked long hours in the New Mexico desert to build a jumping robot, one that could read a map and find its way around inhospitable terrain, deciding on its own whether to move by means of energy-efficient rolling or by energy-demanding hopping. It had to navigate and path-plan, in engineer parlance, making use of complicated algorithms that Salton and his team strung together. The electronics had to be squeezed into a tight space, because the smaller they could make the vehicle, the more fuel-efficient it would be. And it had to be sturdy, partly so the electronics could survive each jarring landing, and partly so it could jump again.

"The real limit is the landing," said Salton, as he dusted off the hopper. He had a drawn, marathoner face above a pressed shirt, khakis, and a sensible pair of black shoes. He had obvious respect, even affection, for it: The old robot had hopped more than 300 times, and, like a coach rationing the playing time of his star forward, the one with the creaky knees and bad back, Salton wanted to limit further impact.

I asked him if it had a name, assuming it didn't.

"Lewis," he said, as he lovingly regarded the robot—the other prototype robot, in another suitcase, was named Clark. "I wanted to name them Lois and Clark—it flies and all that," he explained, "but that was shot down."

In 2010, Salton turned to Lewis for a critical "go/no-go" test before a program manager for DARPA. Lewis successfully executed a series of jumping and wayfaring tasks around the robotic vehicle range, wending its way around cacti, over asphalt, and through windows perched dozens of feet in the air. That evening, to celebrate, Salton and the team went bowling on Taco Tuesday at the Air Force base bowling lanes.

"The secret sauce," he told me—and here I was thinking he was going to say something dark and weapon-y, about how the vehicle had managed to hide a mini–machine gun in its aluminum scales or how vials of plutonium powered the thing—"is how we store and meter the fuel."

Basically, the machine works its way backward each time it must make a jump: By calculating how high it has to jump, it sets aside the right amount of fuel to achieve the task. Then it shoots out an "actuator" (geek-speak for anything that leads to an action), essentially a foot with a foam pad—sort of like the front wheel on R2D2. When the actuator, in a piston action, presses fast against the ground, the hopper springs into the air, tumbling and twisting before coming to a hardy landing on those puffy wheels. It's a tough little guy: The acceleration travels over the four inches of the actuator; the deceleration is spread over the four wheels, each of which has a radius of only about half an inch. "It's an enormous shock; electronics don't like to be shook up at a thousand Gs," Salton says. "Our joke is: Hopping once is easy, because it's mortar." I stared blankly and told him I didn't get the joke, feeling the weight of generations of clueless non-military forefathers pressing upon me. Salton took it in stride: "Oh, it's just that once mortar lands, that's it"—the ordnance has exploded. "Getting the hopper to hop again is the hard part."

We were outside now, in a triangle of asphalt by one of the doublewides, the warehouse, and some outbuildings. It looked like the backlot of a movie studio. A random piece of rocket leaned against a golf cart. A high scaffold stood in one corner, with lights and a set of security cameras, bunched together like grapes, hanging off some of the woodwork. At another corner of the lot, some engineers held aloft gizmos to gauge wind speeds, to decide whether they could launch their hexo-copter, a mini-drone that had six helicopter blades. Nailed to the outside of a shed was a basketball hoop, of course, because the damned things were chasing me.

"Need help with anything?" a bearded post-doc hollered over to Salton as he hunched over a laptop that sat atop a garbage can, in the kind of high-tech/low-tech maneuver that suggested a haphazard, ingenious glue-and-Scotch-tape operation. He was programming some parameters into the hopper's byte-size brain.

Salton un-hunched himself and presented crossed fingers—luck: "Just that kind of help."

He affixed to Lewis a small canister, like the sort you'd find filled with whipped cream; it carried the fuel to shoot the actuator down.

"I feel guilty," he said, and he appeared to be talking to the hopper itself.

"What about?" I asked, as I scribbled down notes.

"I have two very old Labrador retrievers at home, and I feel the same way about this machine as I do about them."

"I see knee problems in its future," laughed a Sandia staffer who was watching.

"It *is* a fairly well-utilized prototype," he admitted.

All told, Lewis cost hundreds of thousands of dollars to build. Unless it goes into the toy market as the most awesome remote-control monster truck ever, it will never be profitable. But as a nimble, lifeless scout, it might save the lives of countless American soldiers.

Out there in the New Mexico desert, I asked Salton what distinguished Lewis from human jumpers. "If you programmed Lewis to do it," I asked, gesturing to the hoop that hung only feet from us, "what would jump more efficiently, Lewis dunking itself or me dunking a ball?" He seemed to think this was an interesting question—Salton actually stroked his chin—leaving me quietly pleased with myself. Or maybe he was trying to figure out how to answer politely a question that was vaguely absurd.

"The human body is not very efficient," he finally said. (Tell me about it, I thought.) "We optimize our gait compared to a robot. But we have torn ligaments. We have banged-up knees. We have hips out of alignment. A lot of the work is wasted on just picking yourself up." The machine humans have engineered to launch itself highest—a space shuttle—has rockets that drop away after they have completed their propulsion work. Human legs, on the other hand, don't detach and float away after they've launched us into the air. "Whether it's biological or mechanical, jumping is a big deal in terms of energy," Salton continued. "It's hard to compete in terms of efficiency with a robot. For us, landing, including the bending of knees to absorb the shock, takes an enormous amount of energy."

Finally, Salton demonstrated the jumping capability of the hopper. With a small remote-control joystick that looked like a run-of-the-mill Nintendo controller, he steered the machine along the asphalt to a 12-foot-high fence. Beyond, desert. He punched a button on his laptop. The actuator pressed into the ground and the hopper went flying, tumbling over itself as it skyed over the fence. Then it landed, hard, on the dirt outside. It was as if it had escaped from a prison. I and the one or two other onlookers cheered and clapped—it was truly amazing, this little thing projecting itself so high up into the air—and Salton steered it in a little circle to make sure it hadn't suffered the electronics version of a concussion. He seemed relieved. I asked him if he was going to command it to jump back. He looked at me like I was crazy before taking his leave to scoop it up in his arms, like a helpless pet.

As I got back in the car that would take me out of the Air Force base, the late afternoon light shining meekly now, I noticed dozens of hopping insects buzzing about us and my thoughts turned back to Malcolm Burrows, in England, with his froghoppers and flying cockroaches. I felt like I had seen both the past and the future of jumping, finely evolved exoskeletons and niftily designed pieces of metal machinery leaping many times higher than their body weight might allow. But they couldn't dance like Michael Jackson, and they couldn't execute a 360-degree jam like Spud Webb. They would never jump up and down, pouting like a child in hysterics, to win a parent's attention. They would not jump with unbounded enthusiasm, like the young Masai warriors, to announce their manhood. They would not jump in lordly leaps, like the dancer Rudolf Nureyev, to seize the affection of another dancer and an entire audience. They had no artistry and they lacked exuberance. They would never jump for joy. They could jump, yes, but they could not express themselves. Perhaps a robot could smoothly leap through a basket. But it could never dunk.

●

Charles is out of time now, and my body is out of energy. I had swatted the ball into the hoop, but these efforts didn't quite have the feel and control—the *snap*—that suggested a true dunk. Charles had thrown the ball up perfectly, tossing it so that it hung in the air like a full moon, and I had, more times than not, gotten my palm onto it. But I hadn't quite gotten all the way on top of the ball—it was just an inch too far—and so I was left slapping the ball toward the basket and hoping it would rattle in. A few did, and a few of these felt right—there had been something of that snappish feeling, and the friction of my sliding hand against the warm, curved metal—but as soon as I landed and twisted toward Charles, I would see him shaking his head. "Asher, Asher, Asher," he'd say. The ball had dripped through the net. "So I didn't dunk it?" I'd ask, sincerely unsure and genuinely hopeful. He'd turn to one of the folks watching—a bearded paunchy guy in glasses, another gym-goer caught up in the hullaballoo—and say, "Was that a dunk?" "Almost," he said, holding his thick thumb and forefinger in the air, a delicate centimeter apart. This stranger, this random witness, casually pinching my potential between his fingers like a bug.

They were turning away, now, the boys, one by one standing up and going back to their kids' basket in the corner of the gym, blithely dunking and horsing around. The moment of possibility had imperceptibly turned to a moment of desperation.

Charles took me aside, to the half-court circle, and, huddling with me, an arm briefly around my shoulder, brandished his iPhone. He had queued up a YouTube clip of his winning jump in the 1991 world track and field championships. "See that burst?" he asked, as if, just by watching, I could absorb his talent. (The frustration that he must endure, the bewilderment, at my inability to do better. And the confidence in himself, after all that hard work, at how very easily it all came. He's in a tough business, it suddenly seemed to me—nothing's as hard to communicate as physical success.) "Look at the explosion of those hips," he said, nodding approvingly as an enduring image of himself flew across the screen, as if he couldn't quite contemplate the gulf between my abilities and his. As if dunking were pure obviousness, and the mystery, if there was one in a mind of such confidence, was why everyone couldn't dunk.

He had a client waiting now. His next appointment. My year was up.

"Just once more," he said for the fifth time; he wanted so badly for me to succeed. And I myself couldn't believe how close I was to accomplishing this thing I had worked on for a year—lacking just that little bit of air, a thimbleful, no more, that could fill the distance between thumb and index finger. I relaxed. I told myself I had already completed the dunk, and that this would be, in Charles-parlance, just play. I smiled. I imagined that thousands of fans were stomping their feet, that cheerleaders waved their pom-poms, that Michael Jordan himself was watching, and that Rebecca, filming all this, was NBC television. "Being in an actual game atmosphere adds adrenaline," Josh Scoggins had once told me, after his first dunk. This was it, the last gasp. I pumped my knees, flew forward, planted hard, drove myself up through the air.

I couldn't quite get my hand over it. Perhaps if I had tried a thousand more times, like a kid mastering a video game, I would have gotten it. But I hadn't. I had failed.

The one or two kids who were still left under the basket quickly got up and grabbed basketballs; they sensed that any lingering would drive home the failure, and for them, at their age, failure was something contagious and best avoided. Terrell started ministering to a client on an exercise bike. "Listen, man, I gotta go," Charles said. "But you keep your head up." I could tell he was disappointed for me. Everyone had turned away. Rebecca was the only one left. "You doing okay, big guy?" she asked as she squeezed my shoulder. She asked me if she should throw me the ball. But my legs were tired— not just from that day's jumps, but from a year's worth of them. Time had run out. I paid my final training receipt, stepped outside, and tugged off my sweaty shirt. I threw my high-tops in the trunk. We drove back to Austin mostly in silence.

Postscript

After my grandmother's death earlier that year, my father asked my brothers and me for advice. At 75 he had grown worn and a little bent. He shuffled. He was keenly aware he might slip into old age. "I'm falling apart" was long his standard line after touching back down from some far-flung academic conference, and we always understood that as a knowing, heroic joke about his boundless energy, the topsy-turvy way ideas were always spilling off of him; now the phrase threatened to carry real freight. With his mother's death, he said he was "unmoored," and I took that to mean he was both directionless and, given the care she required in her last years, liberated. He wanted a new purpose. He wanted to shape up, something that had never been important to him, and, in the way that drives so many of us to exercise, he wanted to recapture a long-ago nimbleness, a physical twinkle. Years before, when I was sick and he was sick in hospital beds half a country apart—he had his prostate surgery in New York on the very day my first week of chemotherapy ended—he wrote me: "Wish we could be having dinner together or going for a walk, you and I. We will."

He was right, of course. Just a few months later, on a late-spring day with the jasmine blossoming, Rebecca and I busied ourselves as we waited for the results of my post-treatment scans. We were at our house, the fixer-upper we had closed on just a week before I had been diagnosed. It was gutted now, the old shiplap and 1920s wall-paper fragments exposed, the framing stripped down to the joists. A real rebuilding job. When the phone rang, Rebecca and I took the call to the porch. "It's not often I get to give patients such good news," said Laurie, the nurse for my oncologist, sounding genuinely happy. We celebrated that night Texas-style, with champagne and barbecue.

My father had also been given the all-clear, and now, these seven years later, I hoped he would find the satisfaction, as I had, of some extra lift in the leg. I used to say if surviving cancer had taught me anything, it was to take pleasure in a full helping of *tagliatelli con funghi*—the not-so-good-for-you creamy pasta dish at the trattoria down the street from my place in Austin. Now, by trying to dunk, I had learned the gratification of discipline, and, with something of the convert's zeal, I began suggesting some of the very exercises, albeit in milder forms, that had been bringing me so close to the rim. He took to them with a schoolboy's diligence. He began walk-ing first one, then three, and finally all ten flights of steps up to his apartment. Maybe not with explosive speed, but he could do it in a steady march. He took brisk strolls in his Manhattan neighborhood. He lost weight, discovering for himself the wonders of yogurt. His posture straightened. He was invigorated.

In a way few of us care to acknowledge, we work out in mod-est defiance of the inevitable. This was, of course, partly at play in my dad's remaking of himself. And it was partly at work in my at-tempt to dunk, made at an age when I was just beginning the slide from peak physical capability to, ultimately, dust. (Why put too fine a point on it?)

At the start of the year, I told myself I was in a race against the imperceptible decline of my physical abilities. This was my last chance to dunk. Yet I discovered that even if we're not all quite dunkers, we have what might fairly be called a superhero's confidence in transformation, a capacity for creative self-improvement that winks at the grave. "I know of no more encouraging fact than the unquestionable ability of man to elevate his life by a conscious endeavor," Thoreau wrote. I think I know what he meant. Even utterly spent, as I pushed through the gym doors early one evening, the sky dark enough that the cars on the freeway had flipped on their lights and you could see the glow of neon popping up here and there from the Action Pawn and the Blazer Tag, I found myself cozying up to the thought that I was lifting myself higher than I had ever thought possible.

I never did slam home that basketball. I sometimes look back at the year of the dunk and wonder why I failed. I had some of the world's best coaches giving me advice and was regularly the sweatiest man in the gym. So I'm left grasping after trivial explanations: Maybe if I had found the right pair of sneakers. Or was it something further back in time, a piling-up of small decisions I had made over many years? Maybe I was just born fated to be earthbound? Whatever the reason, it all added up to an inarguable fact: I couldn't dunk.

My brother-in-law Ben, the basketball player, once wrote about the "climate of assessment" that hovers around competitive athletes: "Childhood, for them, was the game you won at." I'm no professional or college or even high school athlete, so my failure has had no real repercussions—I wasn't cut from the team, I didn't feel lesser than the guy next in line. I was like a lot of people: athletic enough, with a thin desire to win, but never the best and never desperate to be the best. Maybe that explains why, all those years ago, as a carefree boy flailing about the streets of Manhattan, I tried my hand at awning-slapping; unburdened by any expectation of success, I ran along fancy-free. Even now, all these years later, I still occasionally swing away, as if childhood were a game you merely played at. But during the year of the dunk I took those childhood antics suddenly seriously, testing how I might measure up against a sterner yardstick.

Sometimes I wonder what another Asher, whose combination of fortune, effort, and experience had propelled him to dunk on that rim that final day, would be like. Would he have had Charles Austin's natural confidence and easy smile? Would he so trust in his own abilities that the men and women around him would take them for his simple birthright? Zadie Smith wrote about the similar pangs she felt as she ran across a Talking Heads album in a record shop one day. "As I stopped to admire it, I was gripped by melancholy, similar perhaps to the feeling a certain kind of man gets while sitting with his wife on a train platform as a beautiful girl—different in all aspects from his wife—walks by. *There goes my other life.* Is it too late to get into Talking Heads? Do I have the time? What kind of person would I be if I knew this album at all, or well?" The dunk project was my chance to cross the street and walk along that other life. But perhaps, despite my dramatic physical transformation, or because of it, if I had managed to complete that dunk in Charles's gym, it would have signaled no profound change: it wouldn't have been the act of a natural dunker, after all, but the final and fast-fading result of a tremendous effort.

In any case, in the end, that other life, that do-over, remained, like all fantasies, out of reach. Is that the gloomy subtext to my father's rejuvenation? We each can lift ourselves off the ground for so long, but in the final analysis, we're all moving in one direction? I'm not sure which was the more seductive lie spun by Michael Jordan's remarkable hang time: the lie that we could defeat gravity or the lie that time could stand still. Even Michael Jordan had to touch back down eventually.

Months after my final dunk attempt, my APL shoes lie abandoned, thudding about deep in the trunk of my car. A bag of protein-shake mix slumps untouched in a kitchen cabinet, waiting to be thrown away. My gym membership is long canceled. But I still play pickup basketball. I show guys the video of me dunking on our court, and they hold their hands over their open mouths. I can't bring myself to tell them it's less than ten feet, for fear of dashing their own dreams. I'm quicker, partly as a matter of confidence. I take guys off the dribble, or back them up in the post and try a turnaround jumper—my elevation is still better than it was before I began my crazy experiment. I emerged sinewy, tough, and lean. Not wanting to disappoint my mother ("Of course you're always handsome," she tells me on the telephone, "but you're especially handsome now"), my weight remains pretty good. I eat dessert more freely these days and am back to my happy pasta-eating ways. But asceticism clings, like the last streaks of nonfat yogurt sticking to a bowl. I used to be a whole-milk person; now I'm a one-percent kind of guy. I've been known, even now, to decline the services of a bun with my burger. It's a little sad. One evening every couple of weeks, just as the sun starts to set and the grackles muster along the power lines for their evening gabfest, I make my way to the nearby middle school track to do a lonely set of sprints. I put myself through some push-ups and a few sets of sit-ups, lowering myself to the ground with the enthusiasm of a man getting into a cold bath. It's mostly vanity now. This is what happens once you've found yourself with a six-pack, with higher hops, with yet more compliments from your mother. You find it hard to let yourself go. But I'm trying.

Appendix A
The Dunker's Handbook

Suppose, having read this book, you wanted to add inches to your vertical. My sense is that it's hard to lose weight for the sake of losing weight, or to lift for the sake of lifting. Both seem joyless. It helps, instead, to have a purpose, even if that purpose is something as rudimentary as getting yourself to jump higher. The weight loss and the exercise will follow.

Here's a scaled-down version of how to jump higher.

First, measure your current abilities. This'll be crude, but doable: Put a bit of tape on your finger, doubled over into a loop. From a standstill, jump and slap a wall, making sure to stick the tape to the wall at your highest possible point. OK, this is the mark you want to beat.

Jumping, like comedy, comes in threes:

1. Stretching. First, take a light jog, one just long enough that you begin to sweat. You're going to want to improve your explosive capabilities as quickly as you can, so feel free to mix in some high knees, some skipping, some jump rope, as you jog. Now that you're warmed up, you ought to take a solid 10 minutes to stretch. Make sure you get to your quads, your hip flexors, your groin, your calves, your butt, and your Achilles. Bonus points if you stretch your IT band. Think about getting as deep a stretch as you can. Where are you stiffest? Work on more nuanced stretches to ease up these regions of your body. As a reference, I recommend the book *Staying Supple* by the late John Jerome.

2. Lifting. I actually want you to start with no weight at all. In fact, start working out on the very chair you're sitting in right this moment.

Trying standing up and sitting down. Do that a few times, rather quickly. Did you use your hands to help launch yourself up? Now try standing up and sitting down with no hands. Not even on your knees. Just keep them by your sides. Try springing up as soon as your bum touches the chair seat. Can you do that comfortably, 10 times in a row? OK, once you can do that, I want you to do that on one leg: Lower yourself down and pick yourself back up again on the same leg. The other foot shouldn't touch the ground. Hard as hell, right? Get back to me when you can do that 10 times in a row, in three sets. (A tip: Don't use a chair with wheels.)

Another exercise: squats. Stand up. Now drop your bum below the level of your hips. Make sure you keep your heels on the ground. And make sure your knees don't stray ahead of your toes. Got it? (If that's too hard for you, put a board or a folded towel beneath your heels to elevate them slightly.) Do this 10 times, up and down. Each time you go up, go up as fast as you can. Now try doing the same thing while holding a broom over your head, with your elbows more or less locked in place.

Finally, try doing squat jumps. When you explode upward from a squat, follow through and actually leave the ground. Land softly, with knees slightly bent. Lower yourself immediately back into a squat and explode upward again.

Try following these and other leg exercises Monday, Wednesday, and Friday. On Tuesdays and Thursdays, try a track workout. Concentrate on short bursts of speed, starting out no longer than 25 meters and moving to greater distances, stopping at 200 meters. Take a 30-second break after each 25-meter sprint and a one-minute break after each 200-meter sprint. So a workout might look like four 25-meter sprints, three 50-meter sprints, two 100-meter sprints, and one 200-meter sprint. And then work your way back down the pyramid. Or you could do a set of six 200-meter sprints. If you don't have a track available to you—if you live on the nineteenth story of a downtown high-rise—then take the stairs. Try going up two-by-two if you can. Or taking them one at a time as fast as possible, to improve your footwork. Don't use the rail. For a heftier workout, go up two flights, down one flight, up two flights, down one, etc.

Each of these workouts, as well as the leg workouts, should include some abdominal work—sit-ups or planks or whatever works out your tummy.

There are tons of other leg workouts, including ones, like squatting, that involve weights. But for now, these will suffice.

3. Dieting: Eat oatmeal for breakfast. Okay, that's basically your carbs for the day. Find something you like that's low-carb and low-fat and happily snack away at it for the day. For me, this was nonfat yogurt. And I chewed gum like a man desperately trying to quit smoking. I also ate lots of fruit and raw vegetables. Sure, fruit has sugars, but we all learned in elementary school that fruit and vegetables are good for you. No alcohol, I'm afraid. No cookies. No dessert—unless it's fruit. No fruit juices, though. And try to cut down on the dried fruit. Apart from the oatmeal, the yogurt, and the fruit, you ought to stick to protein and vegetables, preferably steamed. Have small helpings of turkey and white-meat chicken; the occasional burger, sans bun; tofu; fish. Try for lean cuts, prepared with little oil. And don't underestimate the incredible edible egg.

Appendix B
The Physics of Spud's Dunk

On Spud Webb fan appreciation night in November 1985, the Atlanta Hawks gave away souvenir stickers instead of the usual posters. "Spud's just too small for a poster," a team official explained. To get to the bottom of Spud's heroic hops—to get a sense of how impressive it is that a man too small to warrant his own fan-night poster exerted enough force to dunk—we can look to two famous laws of physics.*

The first is Newton's second law of motion, known commonly as $F = ma$, and which also can be expressed as $F = m^* dv/dt$, where dv is Spud's change in velocity during the jump and dt is the time over which the jumping force is exerted, i.e., the amount of time Spud spends pushing off the ground before being completely airborne. F is the sum of forces acting on Spud: the force he exerts against the ground when he pushes off to jump, minus the force of gravity, since gravity is a force pulling him in the opposite direction. If we use F_j and F_g to describe the jumping and gravitational forces, respectively, Newton's second law of motion can be rewritten as $F_j = F_g + m^* dv/dt$.

A quick aside: We're constantly balancing the pull of gravity with a force of our own, whether we're sitting down, walking about, or lying down. Gravity pulls us downward with a force of 1G—our weight. To

---

* The physics of even seemingly elementary behavior can get very complicated very quickly. This is a boiled-down look at the physics of the dunk, one that steers clear of certain subtleties that require calculus to explain—and calculus is frankly beyond me.

stand, we must constantly exert an upward force of 1G to counteract it. The net upward and downward force = 0, and so we stay upright. We can express force as a multiple of the force of gravity, and when we do we call them G-Forces, or Gs. (You might notice when you're in an elevator that just as it begins its acceleration upward, you feel momentarily heavier. That's because you're experiencing a balancing force against that acceleration. That is probably between 1G and 2Gs.)

The second famous law at play here is the law of conservation of energy, $mgh = (½)mv^2$, which says all of the energy of motion accumulated during the jump will be used to fight gravity and lift Spud to the top of his jump. Spud's mass, initial airborne velocity, and height gain are m, v, and h, respectively.

We can fill in numbers for all but one of these variables: g is the acceleration due to gravity, which, you may distantly recollect from high school physics class, is 9.8 meters per second squared; Spud's mass during his playing days was 60 kg. Spud could jump at least 42 inches, or 1.1 meters.

Using the law of conservation of energy, we can solve for v, or the velocity Spud needed for a height gain of 42". v = the square root of ($2 *$ 9.8 m/s² * 1.1 m). That comes out to 4.6 m/s. We can also call 4.6 m/s the change in velocity during the jump, dv, since Spud started with zero upward velocity when he planted his feet.

Back to the second law of motion: Doing the math, based on what we just worked out, we learn that the force (in excess of the force of gravity) exerted by Spud Webb when he jumps is (276 kg m/s)/dt.* To express force as a multiple of the force of gravity, i.e., in Gs, we calculate F/(60 kg* 9.8 m/s²) = F/(588 kg m/s²).†

The number of Gs exerted by Spud, then, is actually ~0.5s/dt + 1, or roughly between 3 Gs and 6 Gs, depending on whether that foot-plant lasts a quarter of a second or a tenth of a second. In other words, Webb's acceleration falls somewhere between a fighter jet lifting off an aircraft carrier and a low-to-the-ground F1 automobile cornering a curve at high speed.

Special thanks to Jeff Moses, our family physicist, for walking me through all this.

---

* The 276 figure is 4.6 m/s multiplied by 60 kg.
† The 588 figure is 9.8 m/s multiplied by 60 kg.

Acknowledgments

This book began as a fantasy and grew, in ways I still find hard to fathom, into something remarkably real. Among this project's pleasures, for someone who makes his home in the Texas hinterlands, was catching a peek into that faraway literary world of New York City; I can report that immensely talented and kind people are at work there.

I'm very lucky to have David Halpern as my agent. He showed shrewdness in the early, critical shaping of this book and was a patient, smart advocate as it matured. Thanks, too, to Kathy Robbins, David's partner in crime, and the rest of the Robbins Office for their support.

The irrepressible Vanessa Mobley, bubbly with ideas, felt like a wise sister as much as a sharp editor, setting me straight, in the funniest ways, about not only writing but also the ways of the world. One of her corrective notes: "My friend—no oil on earth is nonfat." Thanks to Claire Potter, who diligently pulled together this book's odds and ends, to Kevin Doughten for his home-stretch work, and to the entire team at Crown—including the art department and Rodrigo Corral for the vivid cover.

It felt like the more I talked about this project, the more real it would be—so thanks to my friends for their indulgence. For their enthusiasms and suggestions, I'm grateful to Nathaniel Mendelsohn, Sebastian Solomon, Marc Bush, Phil Bezanson, Alex Chung, and Julia Markovits; Bob Gee and Carolyn Kimball; and Bill Bishop and Julie Ardery. Gabe and Josh Price and Daniel and Ben Markovits each lent his own brand of valuable brotherly advice.

In important ways, certain people took my idea seriously—arguably more seriously than it deserved!—as it teetered between fantasy and re-ality. Jeff Moses warmed wholeheartedly to the question of the physics of Spud Webb's dunks. Tom Hooven picked out, in his typically inci-sive way, medical specialists for me to approach—without his help, this project might not have gotten off the ground. Along those lines, I'm es-pecially grateful to Steve Doty and Polly de Mille at HSS for their will-ingness to work with a random dude from Texas who wanted to know how to dunk. Thanks, more broadly, to all the people who cooperated in the reporting involved with the book, in particular Charles Austin, Tyler Drake, Malcolm Burrows, Chris Corbett, Kim Geary, Jamie Osmak, Ed Coyle, Jon Salton, Laquan Williams, Josh Scoggins, Mark Goldston, and Todd Wright—and to George Barany, who shared with me the loving family archive dedicated to his parents.

Back in Austin, Debbie Hiott, John Bridges, and Andy Alford at the *American-Statesman* generously granted me the leave necessary to complete this project. Thanks to them and the rest of my colleagues for making the *Statesman* not only a superb newspaper but also such a wonderful place to work—one I have the good fortune to call home.

This book crystallized during my time as a Knight-Bagehot Fellow at Columbia University. The fellowship led me to Jonathan Weiner's il-luminating science journalism class; Jonathan taught us, poetically, to write about scientific research as quanta, or packets of light, a lesson that made a big impression on me. My deep appreciation to the staff of the University of Texas's Stark Center for Physical Culture and Sports, par-ticularly to the avuncular Terry Todd for his help and storytelling on the subject of strength. For their fact-checking help, I'm indebted to Robert Dennison and Holly Duncan.

The much-beloved Jim Phillips served as a crucial reader and con-fidant. One of my joys was meeting Jim at out-of-the-way Austin strip malls for Vietnamese lunch while he held forth on the pleasures of Haskell Wexler or Archer as we went over his squiggles and scrawls. Jim was my first editor at the *Statesman* and remains a mentor, one for whom my fondness is undiluted by the fact that he's a caring ear for so many others. He's that kind of guy.

To my parents, Aimée Brown Price and Monroe Price: It's hard to give thanks for specific things because the truest thing to say would be thank you for absolutely everything.

Finally, my deepest gratitude to Rebecca Markovits. Her influence is on every page of this book. She and I have put together a great life in Austin—a pretty old house with a crazy-faced dog, easygoing friends, dear family nearby. But what's nicest for me is seeing so much of her. She is a partner in the work I do and the life I lead, and, invaluably, she has enough confidence in my abilities to nudge me, in the most thoughtful ways, to do better. This book is dedicated to her. If she were in a dunk contest and I were a judge, I'd tape a crude "1" over the zero and give her an 11. She's that awesome.

Notes

These notes are meant to clarify sourcing when it may not be apparent in the text. I also include links to YouTube videos and some elaboration on points made in the main text.

INTRODUCTION

1 **sellout crowd:** "Spud Webb Wowed Reunion Crowd in '86 Dunk Contest," *Dallas Morning News*, Feb. 11, 2010.

1 **unusually large head:** "Game Changes, but Appeal Remains the Same," *New York Times*, Nov. 20, 1983.

1 **his parents own:** "Spud Webb Wowed Reunion Crowd in '86 Dunk Contest."

1 **three sisters:** "Spud Webb Wowed Reunion Crowd in '86 Dunk Contest."

1 **parking meter:** "Untangling the Webb Mystery," *Chicago Tribune*, Feb. 8, 1986.

1 **can't even palm:** "Spud Webb Wowed Reunion Crowd in '86 Dunk Contest."

2 **Webb makes the league minimum:** "Little Spud Is Big Stuff," *Los Angeles Times*, Feb. 9, 1986.

2 **The winner gets:** "A Speedy Seedling Among the NBA's Tallest Trees," *People*, March 10, 1986.

2 **The judging panel:** "Little Spud Is Big Stuff."

2 **as far back as high school:** "Spud Webb Wowed Reunion Crowd in '86 Dunk Contest."

2 **about the same acceleration as a fighter jet:** A fighter jet accelerates at about 33 meters per second squared, leaving the pilot feeling pushed back in her seat by a force slightly greater than 3 Gs, or three times the force of gravity; put another way, the force exerted on the pilot is about three times the force you'd feel if you jumped off a cliff. For more discussion about the fighter-jet example, see the presentation "Acceleration of Aircraft Carrier Takeoff" at the website of the Khan Academy: https://www.khanacademy.org/science /physics/one-dimensional-motion/kinematic_formulas/v /acceleration-of-aircraft-carrier-takeoff, accessed Aug. 21, 2014.

2 **An avalanche of dunks:** To see these for yourself, check out "1986 Slam Dunk Contest," https://www.youtube.com/watch?v =DnNEe8N1cxs, accessed Aug. 12, 2014.

3 **"In my next life":** Martina Navratilova interview with John Andraisese, Feb. 8, 1986, "1986 NBA Slam Dunk Contest Part ⅔," https:// www.youtube.com/watch?v=jKxwCCSFKPg, 1:30 mark, accessed Aug. 10, 2014.

4 **"God-given talent":** "Quotes of the Week," Associated Press, Feb. 15, 1986.

5 *The Onion:* "97-Year-Old Dies Unaware of Being Violin Prodigy," *The Onion,* Oct. 4, 2010.

5 **"the clock will read 0.00":** Chuck Klosterman, "Is the Fastest Human Ever Already Alive?" *Grantland,* July 16, 2011.

6 **"White men can't jump.":** *White Men Can't Jump,* directed by Ron Shelton, Los Angeles, Calif., Twentieth Century-Fox, 1992.

6 **"I should not talk so much about myself":** Henry David Thoreau, *Walden, or Life in the Woods* (New York: Signet, 1960), 7.

7 **In 2008, candidate Barack Obama:** PierceMedia—610 AM Philadelphia, "Barack Obama on Sports Talk Radio, April 2, 2008," YouTube video, 3:30 mark, https://www.youtube.com/watch?v =M6DT866VXvA.

7 **Obama first dunked:** S. L. Price, "One-on-One with Obama," si.com, Dec. 24, 2007.

9 **The tallest was Danny Rosen:** Danny Rosen is a pseudonym, the only one in this entire book.

9 **"He always kept his poise":** Robert Frost, "Birches," *The Poetry of*

Robert Frost, ed. Edward Connery Lathem (New York: Henry Holt, 1979), 121–22.

CHAPTER 1: ASSEMBLING THE GURUS

13 **Interviews:** Tyler Drake, Daniel Markovits, Polly de Mille, Jamie Osmak, Steve Doty.

15 **did the team snap the streak:** "What's Next Is on Minds at Caltech," *Los Angeles Times*, Jan. 8, 2007.

16 **"I, too, only reached the rim":** Stephen Doty email to author, April 5, 2012.

17 **doctors turned long ago to cadavers:** In the 1980s, a group of researchers conducted the Brussels cadaver analysis. J. P. Clarys et al., "Cadaver Studies and Their Impact on the Understanding of Human Adiposity," *Ergonomics*, Vol. 48, No. 11 (Sept. 2005), 1445–61. Also, Clarys et al., "Human Body Composition: A Review of Adult Dissection Data," *American Journal of Human Biology*, Vol. 11, No. 2 (1999), 167–74.

17 **Her findings:** Polly de Mille, "Asher Price Assessment," Sports Performance and Rehabilitation Department, Hospital for Special Surgery, May 8, 2012.

19 **Michael Jordan's vertical:** http://www.hoopsvibe.com/features/285345 -top-10-vertical-jumpers-in-nba-history, accessed Sept. 3, 2014.

21 **"had no use for girls":** John Bierman and Colin Smith, *Fire in the Night: Wingate of Burman, Ethiopia, and Zion* (New York: Random House, 1999), 23.

22 **"Today I shall be like a god":** Ibid., 22.

22 **"the extra hardships that Wingate had devised":** Ibid., 43.

22 **"The inexperienced soldier":** Ibid., 244.

22 **eulogized him as a "fire-eater":** Ibid., 379.

22 **"all other men seemed uninteresting":** Ibid., 389.

22 **the country's newly minted sport and athletics center:** See the Wingate Institute website, http://www.jewishsports.net/wingate _institute.htm, accessed Nov. 30, 2013.

CHAPTER 2: EVOLUTION AND THE DUNK

25 **Interviews:** Daniel Lieberman, Susan Brooks, Todd Wright, Chris Corbett, Josh Scoggins, Laquan Williams, Demetria Wiley.

25 **During his senior year:** "One Giant Leap," *St. Louis Post-Dispatch*, April 18, 2011.

25 **video of himself dunking:** Jacob Tucker, "Jacob Tucker 2011 Dunk Video," YouTube video, https://www.youtube.com/watch?v=jEgcml1Wx1w, accessed April 16, 2013.

25 **After throwing out:** Ibid.

25 **Under the name "Hops":** "Globetrotter Has a Lot of Hops," *Regina Leader-Post*, April 17, 2012.

26 **the optical device on hand:** "The Longest Jump," *Boston Globe*, Aug. 13, 1991.

26 **"Humans are mediocre runners":** Dennis M. Bramble and Daniel E. Lieberman, "Endurance Running and the Evolution of *Homo*," *Nature* 432 (Nov. 18, 2004), 345–52.

26 **endurance running:** Ibid.

26 **outrun horses over long distances:** Lieberman and Bramble, "The Evolution of Marathon Running Capabilities in Humans," *Sports Medicine*, Vol. 37, No. 4 (2007), 288–90. "In short, for marathon-length distances, humans can outrun almost all other mammals and can sometimes outrun even horses, especially when it is hot."

27 **bonobos have verticals:** M. N. Scholz et al., "Vertical Jumping Performance of Bonobo (*Pan paniscus*) Suggests Superior Muscle Properties," *Proceedings of the Royal Society* 273 (Sept. 7, 2006), 2177–84.

27 **sophisticated projectile technology:** James E. McClellan III and Harold Dorn, *Science and Technology in World History: An Introduction* (Baltimore: Johns Hopkins University Press, 2006), 11.

27 **we could chase our prey to exhaustion:** Scott Carrier, *Running After Antelope* (Washington: Counterpoint, 2001).

28 **"a masochistic lot":** Steven Vogel, *Prime Mover: A Natural History of Muscle* (New York: Norton, 2001), 157.

29 **"a tendency to be barrel shaped":** George Orwell, *Coming Up for Air*, (Harcourt: New York, 1950), 4.

30 **a set number of muscle fibers:** Susan Brooks interview with author, June 4, 2012.

30 **average age of NBA players:** Mike Pesca, "Openly Gay NBA Center 'Happy to Start the Conversation,'" National Public Radio, April 29, 2013.

30 **The clamping closed:** Steve Austad interview with author, May 29, 2012.

30 **Tendon elasticity . . . also narrows with age:** John Jerome, *Staying Supple* (New York: Breakway Books, 1987), 71.

31 **my nerves will fire 5 percent:** Bill McKibben, *Long Distance: A Year of Living Strenuously* (New York: Simon and Schuster, 2000), 43.

32 **"He had a plan":** "The Wright Stuff," *Austin American-Statesman*, March 18, 2009.

32 **"He's part of my family":** Ibid.

32 **he will take home $235,000:** *Texas Tribune* salary database: http://www.texastribune.org/library/data/government-employee -salaries/, accessed July 10, 2013.

36 **"America is the country of young men":** Ralph Waldo Emerson, "Old Age," *Atlantic*, Jan. 1862.

38 **He wore a Daffy Duck tie:** I first made this observation, and a few others that appear in this section, in a personal essay: "Cancer at 26? My Brush with Mortality," *Austin American-Statesman*, Dec. 10, 2006.

38 **The story goes that a biophysicist:** Siddhartha Mukherjee, *The Emperor of All Maladies* (New York: Scribner, 2010), 206. Mukherjee reports, on page 205, that cisplatin "provoked an unremitting nausea, a queasiness of such penetrating force and quality that had rarely been encountered in the history of medicine." Before the advent of anti-nausea drugs, patients in the 1970s treated with the drug vomited, on average, 12 times a day.

CHAPTER 3: THE DUNKING YEAR BEGINS

43 **Interviews:** Steve Austad, Phil Bezanson, Terry Todd, Ben Pollack, Josh Price.

46 **"let us a hear a whistle":** Ralph Waldo Emerson, *Self-Reliance and Other Essays* (Mineola, NY: Dover, 1993), 26.

48 **"the new Professor Dumbbell":** Roberta J. Park, "Healthy, Moral, and Strong: Educational Views of Exercise and Athletics in Nineteenth-Century America," in *Fitness in American Culture: Images of Health, Sport, and the Body, 1830–1940*, ed. Kathryn Grover (Amherst: University of Massachusetts Press, 1990), 131.

48 **"I could have had more wins"**: Bill Bradley lecture, "Values of the Game," delivered Nov. 29, 2012, at the Etter-Harbin Alumni Center Ballroom, University of Texas.

49 **"Never allow others to interfere"**: Timur Tukel, *Air Alert: The Complete Vertical Jump Program* (Charlotte, NC: TMT Sports, 2005), 4.

CHAPTER 4: TAKING THE MEASURE OF THE MAN

54 **Interviews:** Jamal Carter, Luke Anderson, Eric Lougas, Stephen Austin, Sean McKee, Tommy White, James Jackson, Vanessa Streater, Mike Hagen.

55 **Sargent had read *Anatomy, Physiology and Hygiene*:** Dudley Allen Sargent, *An Autobiography* (Philadelphia: Lea & Febiger, 1927), 53.

55 **"To develop my body became an obsession with me"**: Ibid., 54. "The thought that I could grow big and strong under my own tutelage came as a revelation," Sargent writes on page 49 of the book. "I had always felt the joy of existence and the thrill of life that comes from sound health, but I had never interpreted it and directed it. I became suddenly conscious of the physical potentiality for strength and health."

55 **He took up dumbbells:** Ibid., 54.

55 **a "young Hercules"**: Ibid., 61.

55 **"a public prejudice"**: Ibid., 62.

56 **Sargent fled to the circus:** Ibid., 63.

56 **tiresome senior clown:** Ibid., 66.

56 **"the healthy man is the happiest"**: Ibid., 90.

56 **boxing matches:** Ibid., 93.

57 **"make the weak strong"**: Dudley Allen Sargent, "Preparing the Physical Education Teacher," originally a 1908 paper reprinted in *The Making of American Physical Education*, ed. Arthur Weston (New York: Appleton-Century-Crofts, 1962) 183.

57 **"these letters were polite"**: Sargent, *An Autobiography*, 144.

57 **Eventually, one school called upon him:** Ibid., 165.

57 **"to round off the wiry edge"**: William James, *The Gospel of Relaxation*, (New York: Henry Holt and Company, 1899) 53.

58 **"overtired and fagged out"**: Sargent, *An Autobiography*, xv.

58 **a mysterious "unknown equation"**: Dudley Allen Sargent, "The

Physical Test of a Man," *American Physical Education Review*, 1921, Vol. 26, Issue 4, 188 94.

58 **a "vitality coefficient":** Dudley Allen Sargent, "Anthropometric Apparatus with Directions for Measuring and Testing the Principal Physical Characteristics of the Human Body," 1887 (self-published).

61 **1,999 kids worked out:** Sean McKee interview with the author, Feb. 16, 2013.

CHAPTER 5: A NATURAL HISTORY OF LEAPING

68 **Interviews:** Malcolm Burrows, Jeff Moses, Jody Jensen, Charles Austin.

69 **"a novel locking mechanism":** Malcolm Burrows, "Froghopper Insects Leap to New Heights," *Nature* 424, No. 509 (July 31, 2003), 509.

69 **she could clear the Gateway Arch:** "Move Over Flea, There's a Higher Insect Jumper in Town," Associated Press, July 30, 2003.

70 **"postal districts packed like squares of wheat":** Philip Larkin, "The Whitsun Weddings," *The Whitsun Weddings* (New York: Random House, 1964), 24.

71 **a very fine human hair:** "What Is the Nanoscale?" University of Wisconsin-Madison, MRSEC Education Group, http://education. mrsec.wisc.edu/36.htm, accessed July 24, 2014.

71 **An average human being's top reaction time:** Daisy Yuhas, "Speedy Science: How Fast Can You React?" *Scientific American*, May 24, 2012.

74 **"jumping is a self-defeating activity":** Arthur Chapman, *Biomechanical Analysis of Fundamental Human Movements* (Champaign, IL: Human Kinetics, 2008), 146.

74 **"The less ankle flexibility":** Jerome, *Staying Supple*, 13.

75 **get on your bathroom scale:** Chapman, *Biomechanical Analysis of Fundamental Human Movements*, 135; Jeff Moses author interview, Aug. 9, 2014. When you raise your arms above your head, you're lifting your center of mass—even if your feet are not leaving the ground. Essentially you're exerting more force on the ground to lift your center of mass while remaining in contact with the ground—that's what the scale is picking up as you lift your arms. Our weight—as mea-

sured by a scale—is simply an expression of the amount of force we're exerting on the earth. When you drop your arms, you're not working as hard against gravity to stay upright—you're lowering your center of mass. So in that moment in which you drop your arms, the scale lightens.

75 **"Intersegmental dynamics"**: Jody L. Jensen, University of Maryland, 1989.

76 **"bodies are different"**: Correspondence with James Click of the Tampa Rays organization, Feb. 2013.

76 **"various as our several constitutions"**: Thoreau, *Walden*, 12.

77 **"a grew-too-fast kid"**: Richard Hoffer, *Something in the Air: American Passion and Defiance in the 1968 Mexico City Olympics* (New York: Simon and Schuster, 1989), 4.

77 **through with the sport**: Hoffer, *Something in the Air*, 73.

77 **began quietly tinkering**: "The Fosbury Flop: Scorned Method Is Taking Sport to New Heights," *Los Angeles Times*, Aug. 4, 1989.

78 **"Fosbury flops over bar"**: "Amazing Turnabout, Backward Technique, Labeled 'Insane,' Took High Jump to New Heights," *Dallas Morning News*, June 29, 2003.

78 **lifting his center of mass**: "Science of Sport: The High Jump," *Kansas City Star*, Aug. 15, 2008.

79 **sometimes for several minutes**: "Fearless Fosbury Flops to Glory," *New York Times*, Oct. 20, 1968. See also: "Fosbury Flop," YouTube video, https://www.youtube.com/watch?v=Id4W6VA0uLc, accessed Oct. 20, 2013.

79 **the jumpers land on their feet**: Michael Stewart, "Kenyan High School High Jump," YouTube video, https://www.youtube.com/watch?v=-qTAeVGl_e8, accessed Aug. 18, 2014.

79 **"a generation of broken necks"**: David E. Martin, Dwight Stones et al., *The High Jump Book* (Los Altos, CA: Tafnews Press, 1987), 9.

79 **By the Munich Olympics**: Dick Fosbury Training Camp, Bowdoin College, "Dick Fosbury Director Bio," http://www.bowdoin.edu/~pslovens/directorbios.htm, accessed Aug. 14, 2014.

80 **"in a room with a low ceiling"**: Chapman, *Biomechanical Analysis of Fundamental Human Movements*, page 146.

81 **Allex won the state high school high jump**: "San Marcos' Austin Takes High Jump Gold Medal," *Austin American-Statesman*, March 31, 2012.

81 **Charles Austin had grown up poor:** Charles Austin, *Head Games: Life's Greatest Challenge* (Austin: TurnKey Press, 2007), 3.

CHAPTER 6: THE RISE OF THE DUNK

83 **Interviews:** Matt Zeysing.

84 **"You just dunked the ball, man!":** Darryl Dawkins and Charley Rosen, *Chocolate Thunder: The Uncensored Life and Times of the NBA's Original Showman* (Toronto: Sport Media Publishing, 2003), 24.

85 **"jump that rope to death":** Charles Barkley and Roy S. Johnson, *Outrageous!* (New York: Avon Books, 1992), 83.

86 **"The old coach looked like I'd just jumped":** Wilt Chamberlain and David Shaw, *Wilt: Just Like Any Other 7-Foot Black Millionaire Who Lives Next Door* (New York: Macmillan, 1973), 44.

86 **"If I want to see showmanship":** Gena Caponi-Tabery, *Jump for Joy: Jazz, Basketball, and Black Culture in 1930s America* (Amherst: University of Massachusetts Press, 2008), 109. It was also in *Jump for Joy* that I first saw the Zora Neale Hurston quote that opens this book.

86 **"You'd rather look good and lose":** *White Men Can't Jump*, 1992.

86 **After the NBA finally integrated:** Big-time college basketball first integrated in the late 1940s; the NBA saw its first black players take the court in the 1950–51 season.

86 **"subversion of the dominant culture":** Gena Dagel Caponi, *Signifyin(g), Sanctifyin', & Slam Dunking* (Amherst: University of Massachusetts Press, 1999), 6.

87 **"class of incorrigibles":** Interview with Matt Zeysing, June 2012.

88 **"Basketball is for the birds":** Shirley Povich, "Basketball Is for the Birds," *Sports Illustrated*, Dec. 8, 1958.

88 **Allen tried to get the rim pulled up to 12 feet:** "The Leaping Legends of Basketball," *Los Angeles Times*, Feb. 12, 1989.

88 **seize power and express outrage:** Caponi-Tabery, *Jump for Joy*, 106. "The slam dunk returned the locus of power from the owners and coaches to the players," she writes.

89 **save athletes from injury:** "Chronicle of the Jam," *NCAA News*, July 30, 2007.

89 **"if I'd been white":** Kareem Abdul-Jabbar and Peter Knobler, *Giant Steps* (New York: Bantam Books, 1983), 160.

NOTES

89 "the six-foot-two brothers": Pete Axthelm, *The City Game* (Lincoln: University of Nebraska Press, 1970), 127.

89 who called blacks "coons": "Legacy of Rupp Slow to Recede," *Louisville Courier-Journal*, April 2, 1996.

89 "Rupp was so disgusted": Caponi-Tabery, *Jump for Joy*, 108.

90 take four steps at a time: Vincent M. Mallozzi, *Doc: The Rise and Rise of Julius Erving* (Hoboken, NJ: John Wiley and Sons, 2010), 2.

90 "I felt these different things within me": Ibid., 13.

90 By the time he was 14: Ibid., 14

90 "too many nigger boys in it now": Terry Pluto, *Loose Balls: The Short, Wild Life of the American Basketball Association* (New York: Simon and Schuster, 1990), 241.

90 "an act of desperation": Pluto, *Loose Balls*, 25.

91 possessed by a "chocolate thunder": Dawkins and Rosen, *Chocolate Thunder*, 94.

91 started a betting pool: Ibid., 94.

91 A stadium janitor in Detroit: Ibid., 94.

91 "no matter how nice and polite": Ibid., 96.

91 white referees called fouls at a greater rate: Joseph Price and Justin Wolfers, "Discrimination Among NBA Referees," *Quarterly Journal of Economics* 125 (Nov. 2010), 1859–87. Similarly, Northwestern University researchers reported that, in the 2010 NFL season, 65 percent of the players were black but they were flagged for 92 percent of the unsportsmanlike conduct penalties for excessive touchdown celebrations: Erika V. Hall and Robert W. Livingston, "The Hubris Penalty: Biased Responses to 'Celebration' Displays of Black Football Players," *Journal of Experimental Social Psychology*, Vol. 48 (Feb. 2012), 899–904.

91 a column called "The Dunkateer": Dawkins and Rosen, *Chocolate Thunder*, 114.

93 "not a traditional source of pleasure": Philip Roth, *American Pastoral* (New York: Vintage, 1997), 3.

93 "one of the last Jewish infants": Monroe E. Price, *Objects of Remembrance: A Memoir of American Opportunities and Viennese Dreams* (Budapest: Central European University Press, 2009), 1.

95 "Dunking is a power game": Mallozzi, *Doc*, 12.

95 "A dunk is just two points": Ibid., 208.

CHAPTER 8: A DUNK CONTEST

102 **Interviews:** Julius Erving, Darryl Dawkins, Tony Mitchell.

104 **"an infinite number of dunk shots":** Mallozzi, *Doc*, 109.

107 **"that made her ass pop":** Dawkins and Rosen, *Chocolate Thunder*, 59.

108 **"relinquishing doubt and ambiguity and self-inquiry":** Richard Ford, *The Sportswriter* (New York: Vintage, 1995), 60.

112 **dunk off his left foot "in his mind":** "Source: Derrick Rose Cleared to Play," ESPNchicago.com, March 9, 2013.

112 **"I can't dunk":** "Nets' Williams Acknowledges His Limits but Plans to Play," *New York Times*, Feb. 19, 2013.

112 **"My legs feel good":** "Williams Is Soaring Again," *New York Times*, April 22, 2013.

CHAPTER 9: GIRL ON FIRE

114 **"misuses of God's gifts":** Baylor University, "Sexual Misconduct Policy," adopted Jan. 15 2007, http://www.baylor.edu/content/services/document.php?id=39247, accessed Nov. 11, 2013.

115 **her hands are nine inches long:** "Brittney Griner: By the Numbers," PhoenixMercury.com, April 15, 2013, http://www.wnba.com/mercury/news/griner_numbers_130415.html, accessed June 27, 2013.

115 **breaking the girl's nose:** "Baylor's Griner Suspended 2 Games for Punch," Associated Press, March 5, 2010.

117 **Female college basketball players have an average vertical leap:** Jay Hoffman, *Norms for Fitness, Performance and Health* (Champaign, IL: Human Kinetics, 2006), 61.

117 **outjumped 95 percent of the women:** Brian Palmer, "Below the Rim: Why Are There So Few Dunks in Women's Basketball?" Slate, March 23, 2012.

117 **takeoff force actually decreases:** T. E. Hewett et al., "Decrease in Neuromuscular Control About the Knee with Maturation in Female Athletes," *Journal of Bone and Joint Surgery*, Vol. 86, No. 8 (Aug. 2004), 1601–8.

118 **triceps contractions in cats:** E. Henneman et al., "Properties of Motor Units in a Homogeneous Red Muscle (Soleus) of the Cat," *Journal of Neurophysiology* 28 (1965), 71–85.

118 **female muscle tissue:** Suzanne Meth, "Gender Differences in Muscle Morphology," in *Women's Sports Medicine and Rehabilitation*, (Gaithersburg, MD: Aspen Publishers, 2001), ed. Nadya Swedan, 3–4. Also: K. J. Cureton et al., "Muscle Hypertrophy in Men and Women," *Medicine and Science in Sports and Exercise* 20 (1988), 338–44.

119 **dunked 52 times in 32 games:** "Baylor's Griner Becomes 7th Woman to Dunk," Associated Press, Nov. 25, 2009.

119 **"Brittney does what Brittney does":** Imani Stafford at press conference, Feb. 23, 2013.

120 **"soft, mild, pitiful":** Shakespeare, *Henry VI, Part III*, Act I, Scene 4, 140–41.

121 **"someone's child":** "Mulkey Bothered by Social Media Taunts of Griner," Associated Press, April 2, 2012.

121 **an "erotic dimension":** David J. Leonard, "Not Entertained? Brittney Griner Continues to Challenge Expectations," Slam Online, March 1, 2012.

121 **"search my name on Twitter":** "Mulkey Bothered by Social Media Taunts of Griner."

123 **"an unwritten law":** "Griner: No Talking Sexuality at Baylor," ESPN.com, May 27, 2013.

CHAPTER 10: "DAYENU"

124 **Interviews:** Polly de Mille, Jamie Osmak, Alice Price.

127 **"I brag for humanity":** Thoreau, *Walden*, 38.

128 **a Saudi elite soccer player:** Hoffman, *Norms for Fitness, Performance and Health*, 55.

132 **"conspiring against her lone efforts":** Monroe E. Price, *Objects of Remembrance*, 53.

CHAPTER 11: PSYCHING MYSELF UP

134 **Interviews:** Charles Austin, Willy Lenzner, Kim Geary, Josh Price, Nathaniel Mendelsohn, Todd Wright.

135 **very tall Jesus:** Leif Bugge, "1996 Atlanta Olympics, Men, High Jump," YouTube video, https://www.youtube.com/watch?v=y-5lBjj-aS4, accessed April 16, 2013.

140 **"like you fall into step":** Benjamin Markovits, *Leagues Away: A*

First-hand Account of Playing Professional Basketball Overseas (unpublished manuscript), 47.

140 **"ghost images"**: John Jerome, *Sweet Spot in Time* (New York: Simon and Schuster, 1980), 39.

145 **"did not have a brain"**: German Berrios and Rogelio Luque, "Cotard's 'On Hypochondriacal Delusions in a Severe Form of Anxious Melancholia,'" *History of Psychiatry*, Vol. 10, No. 38 (June 1999), 269–78.

145 **"I'm a dead plant"**: A. Vaxevanis and A. Vidalis, "Cotard's Syndrome, a Three-Case Report," *Hippokratia*, Vol. 9, No. 1 (2005), 41–44.

CHAPTER 12: AIMING TOO HIGH

151 **Interviews:** Edward Coyle, Terrell Mercer.

151 **"See those little white marks?"**: Edward Coyle interview, Sept. 2013.

155 **the first time a course in teaching physical education:** Weston, *The Making of American Physical Education*, 55.

155 **"one long orgy of tabulation"**: Harold Rugg, *That Men May Understand: An American in the Long Armistice* (New York: Doubleday, Doran and Company, 1941), 182.

155 **"a modern-day Socrates"**: Steven Horvath and Elizabeth Horvath, *The Harvard Fatigue Laboratory: Its History and Contributions* (Englewood Cliffs, NJ: Prentice-Hall, 1973), 15.

156 **a "scientist's scientist"**: Ibid., 31.

156 **became the first member:** Ibid., 26.

156 **a dog treadmill:** Ibid., 25.

157 **Harvard football players:** Horvath and Horvath, The Harvard Fatigue Laboratory, 62–72.

157 **"the Hobbling Effect"**: G. Edgar Folk, "The Harvard Fatigue Laboratory: Contributions to World War II," *Advances in Physiological Education* 34 (Sept. 2010), 119–27.

157 **"founding father of ergonomics"**: Horvath and Horvath, *The Harvard Fatigue Laboratory*, 73–83. The researchers mentioned here are G. Edgar Folk, Ashton Graybiel, C. Frank Consolazio, and Ross McFarland.

158 **dramatically increased his power output:** Edward F. Coyle, "Improved Muscular Efficiency Displayed as Tour de France Champion Matures," *Journal of Applied Physiology*, Vol. 98 (June 2005), 2191–96. The article's acknowledgments include this now poignant note: "The author very much appreciates the respectful cooperation and positive attitude of Lance Armstrong over the years and through it all."

159 **"hang tough and keep livin' strong":** Lance Armstrong email to author, Jan. 17, 2006.

160 **"No worries at all":** Lance Armstrong email to author, Jan. 20, 2006.

161 **I wrote a personal essay:** I'm referring here to Asher Price, "Me and Lance Armstrong: A Caring Touch in an Hour of Need," Aug. 25, 2012. Portions of that short essay reappear here and there in this chapter.

161 **"I would never beat my wife":** "Allegations Trail Armstrong into Another Stage," *Los Angeles Times*, July 9, 2006.

162 **an error in his calculations:** "Scientific Error Reignites Debate About Armstrong's Past," *New York Times*, Sept. 10, 2008.

CHAPTER 13: SO, CAN WHITE MEN JUMP?

163 **Interviews:** Benjamin Markovkits, Daniel Lieberman, Edward Coyle, Robert Dennison, Terry Todd, Vishy Iyer.

164 **form of expression that embodies:** Caponi-Tabery, *Jump for Joy*, 109.

164 **"you can listen to Jimi":** *White Men Can't Jump*, 1992.

164 **Blacks had a "natural advantage":** "Black Runners 'At an Advantage,'" *Guardian*, Sept. 14, 1995.

165 **"breed his big black":** "Racial Remarks Cause Furor," *New York Times*, Jan. 16, 1988.

165 **"They have different muscles":** "Not-So-Golden Bear; Nicklaus Records His Worst Bogey of All," *Boston Herald*, July 31, 1994.

165 **"biological factors specific to populations":** Jon Entine, *Taboo: Why Black Athletes Dominate Sports and Why We're Afraid to Talk About It* (New York: PublicAffairs, 2000), xi.

165 **"circumscribe possibility":** Entine, *Taboo*, xiii

165 **"strong implications for Jewish comedy genes":** "A Feckless Quest for the Basketball Gene," *New York Times*, April 8, 2000.

166 **"That's a Jewish sport"**: "Nobody Does It Better," *New York Times*, April 16, 2000.

166 **"blacks have something that gives us an edge"**: Entine, *Taboo*, 80.

167 **shoot out tentacles to catch prey**: Vogel, *Prime Mover*, 72.

168 **weighed only 92 pounds**: Michael Bárány and Kate Bárány, "Strife and Hope in the Lives of a Scientific Couple" in *Selected Topics in the History of Biochemistry: Personal Recollections*, Vol. VI, ed. R. Jaenicke and G. Semenza (Amsterdam: Elsevier, 2000), 95.

168 **well-to-do farmer**: Transcript of Michael Bárány interview conducted by Saree Kaminsky for the Shoah Foundation interview on April 11, 1995, http://www.chem.umn.edu/groups/baranygp /michaelbarany/MichaelBáránySr_Shoah_interview_transcript _draft.pdf.

168 **denied admission to university**: Bárány and Bárány, "Strife and Hope," 91.

168 **directed to a train depot**: Ibid., 94.

168 **two-day supply of food**: Ibid., 94.

168 **"transformed the cattle wagon into hell"**: Ibid., 94.

168 **It was Christmas Day**: Shoah Foundation interview.

168 **"lost God"**: Ibid.

168 **slicing some bread**: Bárány and Bárány, "Strife and Hope," 103.

169 **began to study the lives of muscles**: Ibid., 98.

169 **deemed a capitalist**: Shoah Foundation interview.

169 **"escape was not absolutely simple"**: Ibid.

169 **help them escape**: Ted Morgan, "Lord of the Venus Flytrap," *New York Times Magazine*, March 31, 1974, 18.

169 **cow had stepped on a land mine**: George Barany, "The Spirit of Survival: Life Lessons of Holocaust Survivors," memorial speech delivered at Mount Zion Temple, St. Paul, Minn., April 20, 2012.

169 **seven months' pregnant**: Shoah Foundation interview.

169 **through ten miles of snow**: University of Chicago news release, "Michael Bárány, 1921–2011," Aug. 2, 2011.

169 **carried two suitcases**: "Lord of the Venus Flytrap."

170 **"That was a nice walk"**: "The Spirit of Survival."

170 **left their house so early**: "Strife and Hope," 119.

170 **17 good reasons to do push-ups**: Ibid., 135–36.

170 **"inadvertent comparative biochemists"**: Ibid., 122.

171 **cited more than 1,700 times**: Michael Bárány, "ATPase Activity of

Myosin Correlated with Speed of Muscle Shortening," *Journal of General Physiology* (July 1, 1967), 197–218.

171 **"logical and predictable":** "The Spirit of Survival."

171 **straight to graduate school:** "Lord of the Venus Flytrap."

171 **"professors who held hands":** "Michael Bárány, 1921–2011."

172 **between two hefty bookcases:** Jacob L. Krans, "The Sliding Filament Theory of Muscle Contraction," *Nature Education* 3, No. 9 (2010), 66.

172 **muscles that are likely beefsteak-red:** Edward Coyle interview with author, Feb. 2014.

173 **paler, whiter set of muscles:** Ibid.

173 **certain ratio of fast- and slow-twitch:** Robert Dennison correspondence with author, March 2014.

174 **ACTN3 influences the speed:** David Epstein, *The Sports Gene: Inside the Science of Extraordinary Athletic Performance* (New York: Current, 2013), 152–57. Also: D. G. MacArthur and K. N. North, "A Gene for Speed? The Evolution and Function of Alpha-Actinin-3." *Bioessays* 26, No. 7 (July 2004), 786–95.

174 **ACTN3 gene showed up in the top power-oriented athletes:** I. D. Papadimitriou et al., "The ACTN3 Gene in Elite Greek Track and Field Athletes," *International Journal of Sports Medicine* 29, No. 4, (April 2008), 352–55.

174 **significantly slower in a 40-meter sprint:** C. N. Moran et al., "Association Analysis of the ACTN3 R577X Polymorphism and Complex Quantitative Body Composition and Performance Phenotypes in Adolescent Greeks," *European Journal of Human Genetics* 15, No. 1 (Jan. 2007), 88–93.

175 **"misconstrued as rooted in biology":** John Edgar Wideman, "The Architectonics of Fiction," *Callaloo* 13, No. 1 (Winter 1990), 43.

176 **Gabor knew the Nazis would arrest him:** Price, *Objects of Remembrance*, 25.

177 **"it was a feat I was never able to match":** Terry Todd, "Philosophical and Practical Considerations for a 'Strongest Man' Contest," in *Philosophical Reflections on Physical Strength*, eds. Mark A. Holowchak and Tery Todd (Lewiston, NY: Edwin Mellen Press, 2010), 49.

179 **Atlas Sports Genetics began selling:** "Born to Run? Little Ones Get Test for Sports Gene," *New York Times*, Nov. 39, 2008.

179 **failing to prove up its testing:** Alberto Gutierrez of the FDA, letter to 23andMe CEO Ann Wojcicki, Nov. 22, 2013. A March 25, 2014, letter from Gutierrez to Wojcicki termed the firm's "corrective actions" as "adequate."

182 **18 percent of humans who lack:** N. Yang et al., "ACTN3 Genotype Is Associated with Human Elite Athletic Performance." *American Journal of Human Genetics* 73, No. 3 (Sept. 2003), 627–31.

182 **I'm better suited:** A. K. Niemi and K. Majamaa, "Mitochondrial DNA and ACTN3 Genotypes in Finnish Elite Endurance and Sprint Athletes," *European Journal of Human Genetics* 13, No. 8 (Aug. 2005), 965–69.

182 **Michael Jordan's outstretched hand:** Insidehoops.com, "Database of Rare and Interesting NBA Player Measurements."

CHAPTER 14: IT'S GOTTA BE THE SHOES

185 **Interviews:** Mark Goldston, Tim Frank, John Porcari, Cedric X. Bryant, Jesus Moreno.

186 **pitched antiperspirants and hairspray:** "L.A. Gear Names New President, Chief Operating Officer," *Business Wire*, Sept. 26, 1991.

186 **its first drop in earnings:** "Reebok's New Models, Fully Loaded," *New York Times*, Feb. 14, 1989.

186 **shoe you'd wash your car in:** Mark Goldston interview with author, July 2013. (As with much of the material in this chapter.)

186 **brash overachiever:** Ibid.

188 **"They watch you do it":** John Edgar Wideman, "Michael Jordan Leaps the Great Divide," *Esquire* 114 (Nov. 1990), 138–216.

190 **sunny, unseasonably warm February day:** I called the National Weather Service in Atlanta to look up the weather on Feb. 13, 1989, for me. It was sunny and temperatures were above average.

190 **bounded onto a stage:** "Reebok's New Models, Fully Loaded." "We want to make shoes that look great and perform well," Mark Goldston said that day in Atlanta. "'We call it performance-panache.'"

192 **High Priest of High Tech:** Goldston interview with author.

193 **housed, like an Oreo cookie:** Ibid.

194 **"players may not wear any shoe":** "Basketball Shoe Sales Skyrocket After Ban," *Los Angeles Times*, Oct. 29, 2010.

194 "It's the ultimate validation": "NBA Bans Sneakers Made by L.A. Based Company," KTLA News, Oct. 20, 2010.

196 "jumped higher instantly": "Science of Jumpology," http://www .athleticpropulsionlabs.com/technology/technologybasketball/ science-of-jumpology.html, accessed Aug. 1, 2013.

196 "narcissistic" urge to look good: "Marketing Pro Leaves Reebok in the Dust," *USA Today*, Aug. 21, 1989.

CHAPTER 16: CARRYING CAPACITY

201 Interviews: Polly de Mille, Jamie Osmak, Glenn Fleisig, Steve Doty, Benjamin Domb.

202 1999 Israeli national basketball team: Hoffman, *Norms for Fitness, Performance and Health*, 55.

203 one-third of starting pitchers: "Teams Thrown for a Loop by Pitching Injuries," *Boston Globe*, March 23, 2014.

204 7'2" superathlete with freakishly long arms: John Brenkus, *Perfection Point* (New York: HarperCollins, 2010), 143.

204 calculated with 90-percent certainty: Reza Noubary, "What Is the Speed Limit for Men's 100 Meter Dash?" in *Mathematics and Sports*, ed. Joseph A. Gallian (Washington, D.C.: Mathematical Association of America, 2010), 287–94.

205 instances of cosmetic leg lengthening: "New York Man 'Grows' Six Inches Through Surgery," *20/20*, ABC News, Feb. 24, 2012. Steve Doty interview with author, Aug. 2013.

206 "I'm probably at, like, a forty-two": "Rose: Vertical Leap 5 Inches Higher," ESPNchicago.com, Oct. 20, 2013.

206 "Everyone screamed": Josh Scoggins correspondence with author, Sept. 2013.

CHAPTER 17: ALLEZ HOP!

209 Interviews: Charles Austin, Jon Salton, Stephanie Hobby, Philip Heermann.

210 cry of French acrobats: "Miami Heat's LeBron James, Dwayne Wade Revive Art of the Alley-Oop," *Miami Herald*, Feb. 8, 2012.

210 lob play to their springy 6'3" receiver: "Former 49ers Star R. C.

Owens, Known for 'Alley Oop' Catches,' Dies at 78," *San Jose Mercury News*, June 18, 2012.

210 **term took hold in the NBA:** "Chick Hearn: 1916–2002; Lakers Lose Their Voice," *Los Angeles Times*, Aug. 6, 2002.

214 **algae-based fighter-jet fuel:** "UT Bottling Up Potential Fuel Source: Algae," *Austin American-Statesman*, Oct. 25, 2008.

215 **"'I can make a robot do that'":** Sandia National Laboratories news release, "Sandia Hoppers Leapfrog Conventional Wisdom About Robot Mobility," Oct. 17, 2000.

215 **interplanetary exploration:** Ibid.

215 **eight years, a dozen people:** Jon Salton interview with author, Sept. 2013.

POSTSCRIPT

225 **"elevate his life":** Thoreau, *Walden*, 65.

225 **"the game you won at":** Benjamin Markovits, *Playing Days* (London: Faber and Faber, 2010), 233.

226 *"my other life":* Zadie Smith, "Some Notes on Attunement," *New Yorker*, Dec. 17, 2012, 33.

APPENDIX B: THE PHYSICS OF SPUD'S DUNK

233 **"Spud's just too small for a poster":** "A Speedy Seedling Among the NBA's Tallest Trees," *People*, March 10, 1986.

233 **two famous laws of physics:** This passage relies on conversations and email exchanges with Jeff Moses, a research scientist in applied physics at MIT and then a professor of applied and engineering physics at Cornell (and also my brother-in-law), from October 2012 through August 2014.

Index

ABOUT THE AUTHOR

ASHER PRICE grew up in New York City and now lives in Austin with his wife and their dog. He writes about energy and the environment for the *Austin American-Statesman* and plays pickup basketball on his neighborhood court every Sunday morning.